NATIONAL TAXATION
FOR PROPERTY MANAGEMENT
AND VALUATION

NATIONAL TAXATION FOR PROPERTY MANAGEMENT AND VALUATION

Alistair MacLeary

E. & F. N. SPON

An imprint of Chapman and Hall

LONDON • NEW YORK • TOKYO • MELBOURNE • MADRAS

UK	Chapman and Hall, 2–6 Boundary Row, London SE1 8HN
USA	Van Nostrand, Reinhold, 115 5th Avenue, New York NY10003
JAPAN	Chapman and Hall Japan, Thomson Publishing Japan, Hirakawacho Nemoto Building, 7F, 1-7-11 Hirakawa-cho, Chiyoda-ku, Tokyo 102
AUSTRALIA	Chapman and Hall Australia, Thomas Nelson Australia, 102 Dodds Street, South Melbourne, Victoria 3205
INDIA	Chapman and Hall India, R. Seshadri, 32 Second Main Road, CIT East, Madras 600 035

First edition 1991

© 1991 Alistair MacLeary

Typeset in 10/12pt Century Schoolbook by
Rowland Phototypesetting Ltd, Bury St Edmunds, Suffolk
Printed in Great Britain by Page Brothers, Norwich

ISBN 0 419 16820 9 (PB) 0 442 31306 3 (USA)

British Library Cataloguing in Publication Data
MacLeary, A. R. (Alistair R)
  National taxation for property valuation.
  1. Great Britain. Real property. Taxation. Assessment
  I. Title
  336.2220941

ISBN 0-419-15320-9

CONTENTS

Then none was for the party;
Then all were for the state;
Then the great man loved the poor,
And the poor man loved the great;
Then lands were fairly portioned;
Then spoils were fairly sold;
The Romans were like brothers
In the brave days of old.

Lays of Ancient Rome, 'Horatius'

Thomas Babington Macaulay

In a taxing Act one has to look merely at what is clearly said. There is no equity about a tax.

Rowlatt, J. in *Cape Brand Syndicate* v. *IRC* [1972] 1 K.B. 64 at 71

ACKNOWLEDGEMENTS

I am grateful to a number of people who have contributed to the production of this text. Madelaine Metcalfe has been an encouraging and patient editor and Maureen Reid has been a stoically productive and good-natured typist.

My thanks are due to Eilidh Scobbie of the Department of Conveyancing and Professional Practice of Law, and Nantha Nanthakumaran of the Department of Land Economy of the University of Aberdeen, for their expert observations.

Ernest Cruickshank, Chartered Surveyor gave freely of his time to read the draft text and to comment from the practitioner's point of view, and I appreciate the trouble which he took.

Whatever imperfections, inaccuracies or omissions remain they are mine as are the opinions expressed.

The book is dedicated to Norman Harker – who should have written it!

Alistair MacLeary

ABBREVIATIONS

ACT	Advance Corporation Tax
A-G	Attorney General
CAA 1968	Capital Allowances Act 1968
CGT	Capital Gains Tax
CGTA 1979	Capital Gains Tax Act 1979
CUV	Current Use Value
DGT	Development Gains Tax
DLT	Development Land Tax
DLTA 1976	Development Land Tax Act 1976
EC	European Community
FA	Finance Act
F(2)A	Finance (No.2) Act
FYA	First Year Allowance
GDO	General Development Order
HA 1988	Housing Act 1988
IA	Initial Allowance
ICTA 1970	Income and Corporation Taxes Act 1970
ICTA 1988	Income and Corporation Taxes Act 1988
IT	Inheritance Tax
ITA 1984	Inheritance Tax Act 1984
LCA 1967	Land Commission Act 1967
MCT	Mainstream Corporation Tax
para.	Paragraph
PAYE	Pay As You Earn
RICS	Royal Institution of Chartered Surveyors
RPI	Retail Price Index
SA 1891	Stamp Act 1891
S or Sec.	Section
Sch.	Schedule
S/he	She or He
TMA 1970	Taxes Management Act 1970

VAT	Value Added Tax
VATA 1983	Value Added Tax Act 1983
WDA	Writing Down Allowance

Chapter One

INTRODUCTION

SCOPE AND PURPOSE OF THE BOOK

The taxation of property affects us all. People use land and buildings for all their work and leisure pursuits. Most importantly, people need shelter, and in serving that need over 200 000 houses are completed in Britain each year. To fund house purchases financial institutions lend money, and currently the building societies alone hold mortgages to the value of around £160 000 000 000. Property is also the object of commercial finance. About £30 000 000 000 is held in property investments by superannuation funds and pension funds. Property development companies have outstanding commitments of nearly £27 000 000 000 due to commercial banks. The construction industry turns over about £200 000 000 a year. Such figures give some indication of the scale of financial importance of land and new buildings. The initiation of development projects and transactions of interests in land are important in themselves and the effect of tax on holding, developing or trading in property can be critical when decisions are made on land use disposals. But it is not only the large financial institutions and property companies which are concerned about the incidence of tax on property.

For those who are employed in construction, property management or investment or those who find that they have to pay Stamp Duty on the conveyance of their house or Value Added Tax on their bill for hotel accommodation, the incidence of taxation on property is important since it affects their livelihood or their purchasing power. Those so affected include most of us, and most of us know very little about property taxation. People generally only seek advice on particular taxation matters as they need it.

For a relatively small group of people the taxation of land and buildings is a matter of more vital interest. This group will include commercial and professional people whose working lives are intimately connected with the tax system and with the position of land and buildings *vis à vis* the tax system. This group will include lawyers, counsel, accountants, surveyors and civil servants. This group knows about property taxation in considerable detail. It is the very stuff of their lives. They risk their reputations and earn fees by interpreting and practising tax legislation and in representing and advising clients regarding the incidence of tax and tax planning measures. They analyse leading judicial decisions and they prepare quality texts and highly detailed reference manuals. This group knows a very great deal about taxation and some of them are also experienced valuers.

Then there is a group of people who are concerned with the valuation, management, planning and development of land, and for whom the taxation of land is important in so far as it impinges on these activities, but particularly in respect of financial and land valuation implications. This group is made up of surveyors, valuers, estate agents, landowners, property dealers and developers. Some of these may have considerable expertise in land valuation for the purposes of taxation but the majority will command only a basic understanding of the general forms of taxation and the more significant rules in so far as they affect valuations.

This book is designed to be of value to this group of people. It is also intended to be useful to young people aspiring towards professional or academic qualifications in valuation and land management. The treatment of land valuation for the purposes of taxation is a necessary but inconvenient subject in their training or education. Taxation of land does not of itself require any different or unusual valuation methods or techniques. Valuation for taxation is not, of itself, worthy of particular recognition. Equally the knowledge of the taxation system required of the student is usually rudimentary and never likely therefore to form a significant part of any course of study. But basic knowledge is deemed to be essential and the application of land valuation for taxation is often used as an opportunity to inculcate the basic principles and forms of taxation in the student.

Because there is little challenge in either area of wider interest (valuation or taxation), correspondingly scant attention is given to the kind of published material which will be useful to students and practitioners of land management. A single foray into the excellent but voluminous and detailed tomes of taxation which are available is enough to deter even the most inquiring and conscientious of students or practitioners. This publication aims to provide the practitioner with a simplified and concise account of the main features of national taxation and land valuation enabling him/her to refresh existing knowledge and target areas for further more detailed investigation. For the lecturer and student a text is provided which will give a suitable background making prosaic instruction in the tax regime largely unnecessary while providing scope for more detailed concentration on worked examples of valuations and for tutorial discussion. For this objective to be achieved the text has to be one which students will find readable. (This can be a daunting challenge when the subject is taxation!) It is for this reason that the opportunity is taken to develop some wider understanding and theoretical approaches in certain areas. Surveyors and students will wish to see the relationship between taxation policy and practice with other areas of interest such as economics and land use planning.

In attempting to meet its objectives the text is selective in other areas. No aspect of the tax system, or of valuation for taxation is demonstrated exhaustively. There are very substantial omissions. For example the treatment of trusts has been almost totally neglected. The weight of treatment varies between certain areas. The justification for this un-evenness of treatment and lack of detail is that the overriding objectives articulated above require that only those most important matters and points within them can be dealt with while keeping the text at a simple, salient and introductory level.

The reasons for producing a text which intends to meet these objectives, at the present time is, paradoxically, because the taxation system has become simpler in recent years. After ten years of a Conservative administration much has been done to reduce the even greater level of complexity which obtained previously in respect of taxation. Taxation rates have been reduced and harmonized between income and capital. Many impedimenta such as personal reliefs, capital allowances and tax bands have been removed or reduced. There is a clear policy orientation towards taxes on consumption and away from taxes on income. Measures such as the Development Land Tax Act of 1976 have been repealed and others such as Value Added Tax have been extended. Curiously, relief of tax on mortgage interest repayments remains. So far as the entirety of taxation is concerned such changes could be considered to be marginal but they reflect a profound philosophical change which in combination with other government policies encourages individuals and corporations to deal with assets such as land and property in an increasingly liberal and tax-neutral environment, more likely to stimulate enterprise and encourage private led mercantile growth in a supply-side economy. As a corollary, such an approach eschews societal problems and the application of welfare economics and concomitant taxation measures, such as betterment taxation.

These changes are substantially reaching fulfilment at the present time and the application of taxation in respect of all income and capital, including that deriving from land, is sufficiently distinctive to require a complete restatement of the overall position rather than a fragmented update of a relatively large number of different changes. Hopefully therefore the book will be a convenient source of reference which will remain relevant for some time. That having been said there will clearly be changes in detailed matters such as the level of personal reliefs. Some more significant changes may occur. For example, Value Added Tax has been extended to the construction of new commercial buildings in observance of European Community tax harmonization policies. So far as one can see, however, there would appear to be hope for a period of fiscal stability and the taxation measures for land and its valuation should not change significantly therefore from the schemes and methods set out in this text.

This attempt to present a unified picture of the present tax system is also useful since it includes an update within a comprehensive background of the various events and legal decisions which have a bearing on continuing practice in valuation for taxation. As well as outlining the taxation system and highlighting features affecting land and land valuation, the text also includes reference to leading and latest cases so that guidance is given as to judicial thinking in certain areas. Pursuit of this information can often require a determined hunt for information embedded in more highly detailed and comprehensive texts.

The method by which the book seeks to achieve its objectives is firstly to introduce the reader to the broad concepts which lie behind applied taxation measures so that a general understanding of the purposes of taxation and of the British tax system in particular is obtained.

In this section as in the rest of the book it is assumed that the reader has a pre-requisite knowledge of the basic principles of economics and of legal systems. Such knowledge will be available to surveyors and students of land management but the lay reader ought not to have difficulty since care has been taken not to be unnecessarily technical in these areas. However, later parts of the text assume levels of expertise in land valuation and of understanding of planning theory and legislation which may make certain passages or illustrations more obscure for the lay reader.

The text then goes on to examine specific taxation measures. Income and corporation taxes are outlined in Chapter 3 where the more important aspects of income from land are dealt with. Capital allowances, now much diminished in importance as a form of relief from taxation, are dealt with in Chapter 4 and the relevance to property demonstrated. Taxes on expenditure (Value Added Tax and Stamp Duty) so far as they affect land are detailed in Chapter 5. Chapters 6 and 7 cover Inheritance Tax and Capital Gains Tax respectively as the two principal surviving measures for the taxation of capital. As previously, the particular application to property is demonstrated.

Chapter 8 represents a departure from the general form of the book. It deals with the taxation of betterment. Development Gains Tax (Finance Act 1974) and Development Land Tax (Development Land Tax Act 1976) remain collectable on outstanding cases where the charge to tax ocurred before March 1985. However, this is not the justification for the inclusion of this chapter. There is no current betterment tax and outstanding settlements will be increasingly rare.

The reason for the inclusion in Chapter 8 of an extended discussion on betterment taxation is because, firstly, betterment taxes are the only taxes that relate exclusively to land. Secondly, as taxes on development value they are amenable to the expertise of surveyors and valuers who will have knowledge and understanding of planning law and practice, property development and finance, property investment and appropriate valuation techniques.

Students of land taxation and of valuation for the purposes of taxation will benefit from an understanding of the theory and mechanism of attempts to tax betterment. Since the tax is theoretically uncontroversial and since attempts to recoup betterment have had a resilient history over the last forty-odd years, then all of those concerned with land valuation and taxation may find it useful to have a convenient summary of these attempts and some more detailed understanding of the operation of development land tax.

Since a primary purpose of the book is to address the needs and demands that national taxation legislation requires in respect of the value of land and buildings in a wide variety of taxable situations then Chapter 9 devotes special attention to these expectations.

Finally, a partial view is taken of the interaction of taxes in a selected sample of taxable situations with special cognizance being made of land and interests in land together with some of the consequences for land valuation. Here also the opportunity is taken to highlight some of those concepts which are introduced in Chapter 2, such as the need to distinguish between capital and income and investment and dealing. This chapter only attempts to illustrate the taxation consequences of transactions or taxable events and then only to a level of primary consideration. To attempt to be both comprehensive and exhaustive in depth would be inappropriate. It would then be necessary to carefully construct a detailed map through alternative taxation routes under a large number of headings. Such a task is better left to the taxation expert. This author merely wishes to be a guide through the lower slopes, providing a general map of the terrain ahead and who illustrates difficulties and complexities by pointing to one or two minor peaks in the range.

The completeness of this text therefore is not in its coverage or level of detail. It is deficient in both of these qualities. Rather it is in its appropriateness as a source of introductory information to a readership knowledgeable in related fields. The book has been designed for a purpose. It is hoped that those who perceive the purpose may gain most from the text's utilitarian approach.

Chapter Two

THE BRITISH TAX SYSTEM

INTRODUCTION

Before going on to consider particular forms of taxation, it will be useful to have a general understanding of the scope, form and administration of taxes in the United Kingdom.

The taxation system which will be described for the United Kingdom is that which covers England, Wales, Scotland, Northern Ireland and the Scilly Isles, but which excludes the Channel Islands and the Isle of Man.

The taxation system which we shall examine is therefore unique to the United Kingdom. By an important aspect of the Treaty of Rome, particularly articles 95 and 100, the United Kingdom surrenders a modest part of parliamentary sovereignty to the European Community. These provisions are mainly ones designed to prevent discriminatory taxation as within member states. More progressively they give the EEC the power of harmonization of taxation within member states by the use of directives. In this way for example uniformity in the application of Value Added Tax and of Company Taxation is being undertaken. It is also true that the EC itself can raise taxes. EC revenues derive in part from a proportion of VAT, customs duties and agricultural levies collected in the UK.

However, for present practical purposes it can be accepted that tax in the UK can only be imposed by statute by the will of parliament. Equally the expenditure of money by parliament can only be made following a vote on a ministerial financial resolution.

Government spending and the raising of revenues to meet the expenditure therefore requires the sanction of parliament and in particular the House of Commons.

THE PURPOSE AND FORMS OF TAXATION

Clearly in a highly developed economy taxes are required for a very wide variety of purposes. Generally, taxes are needed to pay for items of collective expenditure. For example, the national expenditure on defence, social security or government administration will need to be paid for by taxes. Equally, local expenditure of the same kind which would include education, housing and the administration of planning and estate management services for example would again depend on monies raised through the taxation system to pay for these services.

Taxation can be used as a means of compensatory finance in order to attempt to influence the economy at the macro level. Hence measures of taxation may be used by governments to deal with such key macroeconomic variables as inflation, or the level of unemployment, although 'fine tuning' of the economy along simple Keynesian lines appears to be elusive, particularly in the absence of a good understanding of the effect of taxes on income distribution and economic efficiency.

All such expenditure including government borrowing is met from monies levied through a variety of forms of taxation. National expenditure is met from Income Tax, Corporation Tax, Capital Gains Tax (CGT), Inheritance Tax (IT), Customs and Excise duties, Value Added Tax (VAT), motor duties and so on. Local expenditure is met from a Community Charge together with taxes on business property, (rates) and grants from central government. The sources of tax revenue from the seven years to 1989 are shown in Table 2.1.

It might be supposed that such an impressive collection of taxes was based on some coherent structure evolved over a considerable number of years reflecting accepted principles and sound practice. In fact little is actually known about the benefits of public expenditure and for this and other reasons it is difficult to make sound decisions as to whether a tax base should be predominantly directed at income, wealth or consumption.

Political ideas about equity and wealth distribution vary, and can often be dogmatic. Any observer of tax legislation since say the Second World War would be impressed with the variety of changes which have been implemented, particularly in respect of income or capital gains from property. Efforts to introduce and subsequently dismantle concepts such as the taxation of development rights are extreme examples of this process.

Certainly there are classical principles of taxation. Those which are generally agreed to be appropriate are those which were adumbrated by Adam Smith in 1776. These suggested that a just and equitable taxation system should have the following attributes:

Table 2.1. Inland Revenue and Customs and Excise Duties

Net receipts by Board of Inland Revenue (£ million)

Year	Income Tax	Surtax	Corporation Tax	Capital Gains Tax	Development Land Tax	Estate Duty	Capital Transfer/ Inheritance Tax	Stamp Duties	Petroleum Revenue Tax	Supplementary Petroleum Tax	Total received
1983	31 020	3	5 950	643	68	9	560	1079	5443	676	45 451
1984	32 063	1	7 124	694	75	6	652	956	6882	—	48 453
1985	33 965	—	9 485	855	66	8	846	1159	7369	—	53 753
1986	39 283	—	12 647	971	63	8	939	1700	2698	—	58 309
1987	40 040	—	14 510	1230	31	—	1053	2355	1754	—	160
1988	43 268	—	17 303	2172	17	—	1086	2345	1506	—	67 676
1989	46 540	—	22 436	2002	11	—	1171	2127	1003	—	75 290
1988	61.81	—	29.80	2.66	0.01	—	1.56	2.83	1.33	—	100% Total
1988	36.52	—	17.60	1.57	0.01	—	0.92	1.67	0.79	—	59.08 Gross Total

Net receipts by HM Customs and Excise (£ million)

Year	Value Added Tax	Car Tax	Hydro Carbon Oils	Tobacco	Spirits	Beer	Wine	Cider Wine Perry	Betting and Gaming	Customs Duties	Agri-cultural levies	Ship builders' relief	Other	Total
1983	14 921	683	5497	3750	1310	1648	566	85	613	1095	194	−12	31	30 382
1984	17 060	735	6074	4039	1396	1794	584	93	651	1285	154	−13	27	33 833
1985	19 848	845	6429	4342	1486	1935	623	103	728	1280	142	−15	24	37 770
1986	21 094	927	7174	4585	1505	1968	648	105	763	1266	209	−9	23	40 259
1987	23 118	1081	7674	4818	1514	1989	723	50	825	1449	222	−10	22	43 474
1988	26 620	1332	8503	5021	1591	2084	791	54	893	1644	164	−8	23	48 711
1989	29 290	1554	8739	5155	1562	2104	787	58	956	1795	145	−4	22	52 154
1989	56.16	3.00	16.75	9.88	3.00	4.03	1.51	0.11	1.83	3.44	0.28	0.01	0.04	100% Total
1989	22.98	1.22	6.86	4.04	1.23	1.65	0.62	0.05	0.75	1.41	0.11	−0.003	0.02	40.92 Gross Total

Source: Central Statistical Office.

1. *Equality*, that is that the contribution to tax should be in proportion to the individual taxpayer's income.
2. *Certainty* of amount, time of payment and manner of payment.
3. *Convenience* of payment to the tax payer.
4. *Economy* in the collection of tax.

These might be seen to be sound principles which could therefore stand the test of time. Although modern principles of taxation might also address the benefits to the system which may be gained through qualities such as productivity, elasticity, simplicity and progression of taxes on income. The degree to which any modern tax meets the highest ambitions of these principles varies. Certain of these principles can be in conflict and where there is a relationship between taxes as is common in land taxation then the attempts to avoid double taxation may lead to horrendous complications, rendering impossible the tax payer being afforded any measure of certainty on his liability.

These criticisms should not be taken to suggest that none of these taxes or the tax system is inchoately unfair or unworkable. However, the property owner or property professional should have no doubt that the effects of taxes may not always mirror their intentions, and that it is of paramount necessity to be vigilant in assessing an individual or corporate taxation position in order to avoid the unnecessary payment of tax or to limit the instance of legitimate tax. He cannot assume that the inexorable operation of the taxation system will automatically lead to a just outcome so far as his own taxation position is concerned.

REGULATIONS GOVERNING TAXATION

Taxation is governed by statutes, Acts of parliament, cases decided in the courts, decisions on appeal to the Commissioner of Taxes, 'Concessions' published by the Inland Revenue and practice evolved over the years.

The sources of statute law for taxation are not always easily determined. To be sure the law for any given tax will be found in the legislation which introduced that tax. However, frequently amendments are made to the original act, additional provisions may be introduced and consolidating legislation may be used to contain the law in a single statute.

Very broadly the principal statutes for land taxation will include the Stamp Act 1891, The Taxes Management Act 1970, The Development Land Tax Act 1976, Capital Gains Tax Act 1979, the Value Added Tax Act 1983, the Inheritance Tax Act 1984 and the Income and Corporation Taxes Act 1988. The law incorporated in these and other statutes is amended by Finance Acts each year. Income Tax being an annual tax is subject to review by Parliament in the Finance Bill and the Finance Bill will also be used to alter rules for other taxes.

Clearly, therefore, reference to salient legislation for land taxation purposes is not an easy matter although there are a variety of excellent references for source materials available which are kept up to date either by annual publication or by regular amendment on a looseleaf basis.

STATUTORY INTERPRETATION

Since tax can only be levied by statute, and since in the absence of statute a citizen is not liable to pay any tax, then it follows that he is not liable to tax unless he is unequivocally identified as being liable to tax by the terms of the act which makes the tax lawful.

This has led to the principle of strict interpretation of statute. This can benefit the tax payer if there is any ambiguity as to his liability to tax when he may enjoy the benefit of doubt. However, this is a double-edged sword in so far as it also leaves open the possibility of literal interpretation of statutory provisions which may cause originally unintended hardship to the tax payer.[1] Literal interpretation also tends to militate against ingenious and imaginative interpretation. However, the courts would appear to be entitled to depart from a literal interpretation if the outcome otherwise would be patently absurd.[2] The principle that a person coming within the letter of the law must be taxed however great the hardship together with the corollary of the principle that the person from whom the state sought payment of tax cannot be taxed unless he can be brought within the specific letter of the law now appears to be much modified by the cases of W. T. Ramsay v IRC.[3] Eilbeck v Rawling,[4] Burmah Oil Co Ltd v IRC,[5] and Furniss and Dawson.[6] From the decisions in those cases it now seems clear that as well as examining the incidence of taxation within one step of a series of related transactions the structure of the whole transaction may be reviewed. As a consequence, while it continues to be quite legitimate to manage commercial transactions in a manner that is tax efficient it is not acceptable to introduce steps into a transaction the sole purpose of which is to arrange for the avoidance (not necessarily the evasion) of tax. Furniss and Dawson, and the other cases, have made it extremely difficult to pursue blatant strategies of avoidance.

Since taxes acts apply to the whole of the country then they are so far as possible interpreted to achieve a similar effect throughout the whole of the UK, and for this reason decisions made in any of the countries comprising the UK will be accepted in another country within the UK.

Where there may be doubt regarding the interpretation of tax law then the normal procedure is for a tax payer to argue the point by debating the issue with the local Inspector of Taxes. If as a result there is no agreement regarding interpretation then either party can appeal to the General or Special Commissioners. If the ruling of the Commissioners on a point of law is not accepted then an appeal can be made to the Courts. Clearly the decisions particularly of Special Commissioners can be of value in interpreting taxation law. Until the present time the Inland Revenue have declined the invitation to publish decisions of Special Commissioners although there is considerable and increasing pressure for them to do so.

On the other hand the Inland Revenue is helpful in producing statements of practice. These make a useful contribution towards the practical administration of the taxation system. There are two different kinds of statements: 'Concessions' and 'Statements of Practice'.

In addition the Inland Revenue produces 'Advance Rulings' outlining the likely tax consequences of a particular form of transaction resulting in tax liability. Advance Rulings are not generally available although practical advice may be provided in respect of a detailed proposal.

ORGANIZATION OF THE INLAND REVENUE

The administration of the tax system in the UK is the responsibility of the Board of Inland Revenue, the members of which are called The Commissioners of Taxes. The Board is responsible to the Treasury and is responsible for the collection of income Tax, CGT, Corporation Tax, and DLT (TMA 1970 Section 1). The collection of Inheritance Tax is separately administered by the Capital Taxes Office (ITA 1984, Part VII).

Responsible to the Board are Inspectors of Taxes. Inspectors assess taxes on the basis of information received. Tax payers are under a statutory duty to provide a return of income (TMA 1970 s.7). This information enables the Inspector to make an assessment although if the information contained in the return is unsatisfactory then the Inspector can make an assessment using his own judgement. (TMA 1970 s.29, see also Donnelly v Platten).[7] Inspectors also issue code numbers under PAYE, negotiate adjustments in respect of assessments and authorize repayment where appropriate.

Inspectors can have recourse to other sources of information. They have the power to examine the documents of the tax payer and the Finance Acts 1976 and 1989 extends this power to enter and search premises and persons on the premises in order to acquire evidence where there is suspicion of a tax fraud. The Taxes Management Act 1970 (ss.61 and 68) provides the Inland Revenue with the means of collecting taxes.

The responsibility for the collection of taxes lies with the Collector of Taxes who has the power to distrain on a tax payer's furniture and goods and to sue if he is in default.

RETURNS OF INCOME (TMA 1970 ss.7–9)

As has been mentioned above it is the duty of every person who may be chargeable to income tax to send a return stating the composition and amount of that income each year to the Inspector of Taxes.

The normal procedure is for the taxation office to deliver a standard form for completion and return on which the taxpayer can detail his profits and gains for the preceding year. Should the taxpayer fail to make a return of income within one year of the end of the year of assessment then he will be legally liable for penalties. The failure of a taxpayer to inform the tax office of a liability to charge can mean a fine equal to the income due on the undisclosed income (FA 1988, s.120).

Although a tax form will usually require return within thirty days this may not always be practicable, but it is still advisable to make returns in sufficient time for the Inspector of Taxes to issue a notice of assessment which is normally by the end of the calendar year. This will enable interest on outstanding balances due to be avoided.

The standard form issued by the Inspector of Taxes conveniently sets out under different headings where separate sources of income may be entered as well as providing sections where claims for allowances or reliefs from tax may be made.

When returning the form the taxpayer has to sign a declaration confirming that the information provided is both correct and complete.

In the case of partnerships required to declare income then there is a special form which allows for one of the partners to complete information on behalf of both her/himself and her/his partners.

For a trading operation it is usual to submit a Balance Sheet and Profit and Loss Account (to the Inspector of Taxes) which have been audited by a professional accountant. In addition the accountant may submit taxation liability calculations together with supporting evidence. In this way the assessment of tax of the business may be agreed with the Inspector of Taxes before the formal return is made thus saving time and effort for both parties.

NOTICE OF ASSESSMENT (TMA 1970 ss.29, 30, 34–38 and ICTA 1988 s.2)

During the fiscal year, and usually at the end of the calendar year, notices of assessment are forwarded by the Inspector of Taxes which show the amount which is assessed as being due together with a note of any entitlement to allowances and reliefs and the consequential net amount of the tax payable. If a 'notice to pay' accompanies the notice of assessment then the tax is due to be paid thirty days after the issue of the notice (ICTA 1988 s.4).

Where tax is deducted from the wages and salaries of employees by their employers under Schedule E (see Chapter 3) then the notice of assessment is forwarded to the taxpayer after the year of assessment which terminates on the 5th April. This notice of assessment is thus able to show how much tax was due for the year of assessment based on the employee's total income as well as showing how much tax has already been paid by means of deduction from his or her emolument. An indication can then be given of any tax over (or under) paid and the balancing amount settled.

In normal circumstances an assessment will be based on information received by the Inspector of Taxes and provided on the returns of income provided to him. The Inspector is free to make an assessment of tax due if he is not satisfied with the information provided to him.

If the Inspector discovers that income has not been assessed or has been insufficiently assessed, or that excessive reliefs or allowances have been given, a new or further assessment can be made to rectify the matter. It is clearly a matter of some importance to the taxpayer to know what latitude the Inspector may have to 'discover' information of this kind. The judgement in R. v. Kensington Commissioners *ex parte* Aramayo[8] suggests that whenever it appears as new information that the taxpayer has been undercharged then there is a 'discovery'. The latitude afforded to the Inspector appears to be very wide therefore. If an error or mistake has been made in respect of assessment and subsequently discovered by the taxpayer then application may be made to the Commissioners for relief (TMA 1970 s.33). However, the claim must be made within six years after the end of the period of assessment within which the mistake was made.

APPEALS

The tax payer has a right of appeal against an assessment provided he gives due notice within thirty days of its issue (TMA 1970 s.31). The appeal must be made in writing to the Inspector and must state the grounds of appeal. Generally, the appeal would be made to the General Commissioners of Taxes who are lay persons of local repute, such as magistrates, quite independent of the Inland Revenue. Their function is to hear appeals by the tax payer from the decision of the Inspector and to protect the interests of the tax payer subject to the law. Alternatively the tax payer at his choice can appeal to the Special Commissioners.[9] These are professionals who are experts in taxation law and practice. As well as having specialist expertise they generally have more time available than lay General Commissioners to devote to the consideration of a particularly complicated case. On the other hand the General Commissioners who sit locally are usually more accessible and will deal with the matter more quickly and no doubt much more cheaply. This may be particularly important where costs are not awarded and legal aid is not available.

It should be remembered that in certain instances the legislation specifies whether the appeal will be to the General or Special Commissioners. Since the Finance Act 1984 the General Commissioners may refer a complex case to the Special Commissioners without having to receive the permission of either party although, in that event, the General Commissioners must take into account any representations made by the parties. The opportunity of appeal may also lie to the Lands Tribunal.

Clearly, the Inspector himself may refer an appeal to the Commissioners but the procedure of appeal once commenced can only be terminated with the consent of the Revenue (TMA 1970 s.54).

At appeal the burden of proof is on the tax payer to show that the assessment made by the Inspector is incorrect.[10]

If the tax payer does not agree with the decision of the Commissioners on a point of law, but not on a point of fact, he can then pursue his appeal to the courts by way of a stated case (In England the High Court, in Scotland the Court of Session and in Northern Ireland the Court of Appeal (N. Ireland).

Finally, and still on a question of law, a further appeal from the Courts to the House of Lords is available.

NOTES: CHAPTER TWO – THE BRITISH TAX SYSTEM

1. *Vesty* v. *IRC* [1980] S.T.C. 10 and *Cape Brand Syndicate* v. *IRC* [1921] 1 K.B. 64.
2. *A-G* v. *Hallet* [1957] 2 H. and N. 368.
3. [1981] All E.R. 865.
4. [1981] S.T.C. 174.
5. [1980] S.T.C. 731.
6. [1984] A.C. 474.
7. [1981] S.T.C. 504.
8. [1916].
9. In Northern Ireland there are no General Commissioners and appeals are therefore made to the Special Commissioners. An appellant has the right, however, to elect for a hearing in the County Court.
10. Norman and Golder [1945] All E.R. 352.

Chapter Three

TAXATION OF INCOME

INCOME TAX

Income Tax has a long history in the UK, having been introduced in the first instance during the Napoleonic Wars when 10% of gross income was levied in order to raise finance to prosecute the European War. Income Tax only became a permanent feature of British tax structure in 1842, when it was reintroduced at the standard rate of 7d in the £ (i.e. 3%). The highest rates of tax have been reached during times of war (Crimean War 7%, 1914–18 War 30%, Second World War 50%); currently the basic rate for Income Tax in the UK is 25p in the £.

Until this century, however, liability to Income Tax was a burden carried by a small minority of the population. The total number of tax payers did not exceed 1 million until the twentieth century. It was not until 1907 that a distinction was made between earned and unearned income, and it was not until 1909 that taxation became progressive. In that year Lloyd George introduced a budget which imposed a 'super-tax' on incomes over £5000 per annum. In 1928 'super-tax' was renamed 'surtax' and in 1973 'surtax' was abolished. Unearned income surcharge payable at 15% on such income received was abolished in 1984 (when the limit was £7000), so that today Income Tax is levied as a single graduated personal tax.

Of all the direct taxes Income Tax is the most important (see Table 2.1). The majority of working people now come within the ambit of the tax, the effective incidence of Income Tax therefore heightens the awareness of taxation and government spending. There are other advantages. For government there is the benefit that as incomes rise it can collect more money and also by establishing tax thresholds the government can influence important factors in the economy such as the propensity to save.

Virtually all adults in full-time working in the UK contribute to Income Tax (i.e. all those who earn more than the basic personal relief, which is £3005 in 1990/91). Those who do not pay Income Tax directly will almost certainly be dependent upon an income tax payer, in receipt of tax exempt income (such as student grants) or living on a very low pension.

Income Tax as a tax on earnings is seen to have a number of advantages. Provided the rates are not excessively high it appears to have a smaller disincentive effect than other taxes and is an effective tax in realizing the goal of redistributing income. However, these general truths can be disputed. Arguments can be led to suggest that there is a decline in managerial morale related to a high incidence of personal taxation, and at the other end of the scale there is the phenomenon of the poverty trap where those paying tax but receiving other benefits may find themselves worse off (through increased tax and lost benefits) if they earn one extra £1 per week.

FORMS OF DIRECT TAXATION

The structure of Income Tax in the UK comprises some of the characteristics of each of the principal systems of direct taxation which are theoretically postulated.

Comprehensive income tax

The first of these is Comprehensive Income Tax which is based on the net depletion principle. The idea of this system of direct taxation is that any individual should be taxed on all forms of income (i.e. comprehensive income) which can be broadly defined as the amount which he consumes without depleting wealth.

Such a system requires the taxation of capital gains at Income Tax rates. Since it is relatively easy to turn investment income into capital gains there is a strong argument in favour of this approach. It is the practical difficulties which inhibit its introduction, since it will require the valuation of all assets on an annual basis. Tax under this system would be payable on capital gains as an increase in wealth whether that capital gain had been realized or not (although presumably capital losses would be deductible). Clearly this carries the risk that in any particular year or years the owner of a large amount of capital might be assessed as having a zero or negative comprehensive income although retaining the ownership of a large absolute amount of wealth.

Imputed income would also necessarily have to be part of the logic of a Comprehensive Income Tax. Imputed income arises where an individual owns an asset with potential for income production but consumes such benefit himself.

A relevant example of imputed income is a tax on the residence of the tax payer. Indeed such a tax used to be levied in the UK under Schedule A (infra). Under Comprehensive Income Tax imputed income would have to be reintroduced on houses and would have to extend to other similar valuable items.

Expenditure tax

Since Income Tax is so widespread and paid by such a large proportion of the population it becomes the least popular direct tax. Alternatives are therefore always of interest and a much vaunted system of alternative taxation is based on the idea of direct taxation on expenditure.

The principle here is that the tax is based on an individual's consumption expenditure. However, this is a different concept from indirect taxes on expenditure with which we may be familiar, such as Value Added Tax. The individual would be taxed on the total resulting from his annual receipts less amounts spent on purchasing, plus amounts drawn from capital and spent on consumption.

An expenditure tax avoids many of the practical problems associated with a Comprehensive Income Tax. For example it is not necessary to value wealth on an annual basis. It is not necessary to index for inflation or allow for depreciation, and since it has the attraction of removing any artificial distinction between capital and income it will encourage saving.

However, practical difficulties remain, particularly in respect of the means of determining income and of defining expenditure, as well as the short-term but real problems of arranging a transfer from the present system to an expenditure tax system.

UK INCOME TAX

Although it contains elements of these two systems, UK Income Tax is not on the other hand a classical case of British compromise. Income Tax in the UK is clearly based on certain well-established principles. For example, Income Tax may be levied progressively so as to fall proportionately more heavily on those with larger incomes.

Again saving is encouraged by tax incentives. In the particular case of property, relief from taxation is available for saving by way of investment in private housing.

However, there are structural difficulties in the system simply because it has not been designed from first principles. Hence, traditionally capital gains in the UK were treated as windfall profits and not taxed, now that they are taxed they are treated quite separately from income, a matter which tends to militate against the general acceptability of theory. The system has recently reduced this difficulty by harmonizing rates of taxes from capital gains and income although the rules and reliefs remain very different.

Income chargeable

Income Tax is imposed for each year of assessment (6 April to 5 April). Income Tax is charged at basic and higher rates. Currently, the basic rate is 25% with a higher rate of 40%.

An individual's actual tax liability is modified by means of personal allowances (infra). Section 32 of the Finance Act 1988 provides for a new system of taxing married couples which has operated from 6 April 1990. Each spouse is taxed separately on both income and capital gains. Each has a single person's relief and each enjoys full capital gains tax annual exemption. There is a transferable married couple's allowance, which is given primarily to the husband and only passes to the wife if he has inadequate income to use it.

The profits of limited companies and other corporate organizations are subject to Corporation Tax (infra) and not Income Tax.

Tax schedules

There is no formal definition of income and even a theoretical definition can be difficult. However, Section 1 of the Taxes Act 1970 refers specifically to income being taxable if it falls within one of the schedules of the Act.[1] Quite clearly this means that if income does not fall to be specified within one of these schedules then it cannot be taxed. A brief description of these schedules follows.

Schedule A (ICTA 1988 s.15)
Schedule A taxes income from land and buildings. This may be in respect of rent (from a weekly tenancy to a long-term ground rent) or to premiums receivable on the grant of a lease (for a term not exceeding fifty years). The actual tax payable is calculated on the net income receivable after deducting allowable expenses from gross rents on an annual basis.

Prior to 1963 there was a charge to tax under Schedule A related to the annual value attributable to the beneficial ownership of heritable property.

Schedule B (ICTA 1988 s.16)
Under this Schedule income from woodlands managed for profit was charged to tax. The amount of income assessed for tax was assumed to be

one third of the annual value of the land if let in its natural and unimproved state. This Schedule was abolished as from 6 April 1988 (FA 1988 s.65 and Sch.6).

Schedule C (ICTA 1988 s.17)

This Schedule levies tax on interest annuities and dividends paid in the UK out of public funds. The payments concerned may be made by a foreign government, public authority or institution as well as the UK government.

The basic rate assessment is made on the paying agent and since it is deducted at source does not directly affect the ordinary tax payer.

Schedule D (ICTA 1988 s.18)

This Schedule provides for the taxation of annual profits or gains which fall to be considered under one or other of 6 sub-divisions or 'cases'.

Case I

Case I taxes the profits of a trade or business (other than a business covered by Case II). The taxable profits of a business are assessed to tax quite independently of any liability to tax of the person who owns the business and who is separately assessed. The profits from the trading operations of the business are assessed under Case I but profits from any other source might be liable to tax under one or other of the Schedules or another case within Schedule D.

Where a business is unincorporated (i.e., owned by one person or a partnership) the reliefs and the allowances applicable to the owners are usually set off against the assessment of the business before the actual tax is calculated and the owner, or partnership assessed under Schedule D Case II). On the other hand a company having limited liability is a legal persona and the tax assessment made on profits from trade or business will not be subject to any of the reliefs which might be due to the shareholders. This will be true even if the limited company is owned by one person only.

In the case of a partnership comprising a number of owners then a tax return must be made out in respect of the business or firm with the partners making an individual return in respect of their own incomes including that received in a partnership. The tax chargeable on the firm is then apportioned to the individual partners.

Case II

This case applies to the taxation of a profession or vocation (for example lawyers, doctors or surveyors) where they may be operating a sole business or partnership.

(Note: If such professionals are employed they are assessed under Schedule E – see below).

Case III

This case taxes interest, annuities and other annual payments receivable by the tax payer from loans including receipts from public revenue dividends where they do not otherwise fall within Schedule C; examples of this kind of income might include government securities where tax is not deducted at source (e.g., 3.5% War Stock), cooperative society dividends or bank deposit interest.

Case IV

Assessments under this case are in respect of income from securities outside the UK for example receipts from foreign securities (and not included in Schedule C).

Case V

This case taxes the profits or gains arising from possessions held outside the UK. For example, included in this case might be dividends from overseas shareholdings or rents from properties abroad.

Case VI

Case VI is designed to cover any profit or gain which might not otherwise be identified by other schedules or cases. The kinds of income which might be covered by Case VI would include for example profits from furnished lettings or development gains (FA 1974).

Schedule E (ICTA 1988 s.19)

Schedule E taxes income from wages, salaries, bonuses, commissions, fees and benefits received in connection with an office or employment.

This schedule has three cases which refer specifically to the location of employment, so that as well as UK residents receiving emoluments in the UK, non-residents who are employed in the UK or residents in the UK receiving remuneration from outside the country are also specified as being eligible for assessment under the schedule.

Schedule F (ICTA 1988 s.20)

This Schedule covers the income tax on dividends and other distributions payable by companies resident in the UK. Taxes due on the dividends of limited companies or other corporate bodies are assessed on the dividend received plus a 'credit' representing tax 'imputed' to the dividend (see infra). Effectively the basic rate tax is deducted at source.

Taxes on income from land and building

For the purposes of assessing tax due on income from land and buildings, therefore, the schedules which are particularly important are A, B and D (Cases I, V and VI). These will be dealt with later.

Timing

The question of when an income is received and when tax may be payable is clearly important. As a general rule an assessment to tax cannot be made until a payment is received. Equally it is also generally the case that payment received will be treated as income relevant to the period when it became due.

Although it is normal for income to be assessed within the financial year that it accrued, i.e., on a current year basis, it is also the case that in the UK some sources of income are taxed with reference to the year prior to assessment, i.e., the preceding year basis. In this case it is felt that the sources taxed as income throughout the year of assessment may not vary, in which case the preceding year's income represents a measure of the income for the year of actual assessment. Clearly where this occurs income will be taxed at the rates in force in the year of assessment, not those obtaining in the year when the income was actually received.

If there is a substantial fluctuation in the income arising from any particular source between periods then the situation can arise where the assessment of tax in one year may be quite out of proportion to the income actually received in the same year. This has been a particular problem for farmers, who may now elect to average their income.

Income not subject to tax

Given the apparently exhaustive nature of the schedules in the Income and Corporation Taxes Act 1988 including the miscellaneous Case VI of Schedule D, it might be thought that there is very little prospect of any income not being liable to assessment of tax.

However, the courts have construed the use of the term 'annual profits' in a way which permits profits to be assessed as income only if they are recurrent.[2]

As we have seen already the courts choose to draw a limiting view of income by making a distinction between income and capital. This basic distinction between income and capital appears to be a cardinal feature of the UK taxation system. This has meant that specific legislation has had to be introduced to deal with, for example, taxation on premiums for leases.

Whereas non-taxable receipts do exist as a consequence of such interpretation such income is generally at the margin. Although important areas are free of Income Tax, such as winnings accruing from gambling, loans, and most gifts, some like capital gains are taxable under headings other than Income Tax.

Personal allowances (ICTA 1988 s.257)

Since the UK system of Income Tax is applied on a universal basis it does not attempt to provide any individual reliefs through the provision of very low rates of tax or by means, say, of an automatic exemption to the first slice of taxable income across the board.

Rather any adjustments to income taxation assessment is carried out by allowing relief specific to individuals. Personal relief is used to achieve as equitable as possible a distribution of the tax burden according to individual circumstances. Hence in the individual case certain income from social security payments would be exempted from Income Tax. Reliefs generally would afford the greatest comfort to those on lower incomes.

The normal tax unit is the individual although to a limited extent the tax unit can embrace a married couple (through the transferrable married couples allowance) or dependent children (whereby income derived from funds provided by a parent, paid to an unmarried minor child, is taxed as partner's income).

Personal reliefs are allowances which may be deducted from total income in order to calculate taxable income. However, they can only be claimed by individuals. A partnership will be taxed for income in the normal way and will not of itself attract personal reliefs. Such personal reliefs can only accrue to individual partners of the partnership. Reliefs cannot be rolled forward or backwards in time, they may only be used in the particular year of assessment and not another year, and as a general rule may only be claimed by residents of the UK.

(a) Personal Allowance (ICTA 1988 s.257)
This relief on taxation is granted to the individual. Section 257 of ICTA 1988 affords a personal relief (currently £3055 – FA 1990).

As from tax year 1990–91 when husbands and wives are taxed as separate individuals each has her or his personal relief. A married couple have the benefit of a married couple's allowance, which will be given to the husband but will be assignable to the wife if he is unable to use it, (currently £1720).

(b) Age Allowance (ICTA 1988 s.257)
For individuals aged sixty-five or over and whose income is not in excess of £11 400 then a single person may claim an age allowance £2300. In the same circumstances a marriage allowance amounts to a larger sum

of £5815. This allowance is in substitution for the ordinary personal allowance. There is an increased age allowance for people over seventy-five.

In circumstances where the income may exceed £12 300 the extra allowance is progressively reduced.

(c) Additional Personal Allowance (ICTA 1988 s.259)

This allowance is particularly intended to benefit single parent families. A deduction of £1720 is allowed in respect of qualifying children resident with the parent if that parent is (1) a married man whose wife is totally incapacitated, (2) any person other than a married man who has a qualifying child living with them for the whole or part of the year.

For the child to be a qualifying child then the child must be under the age of sixteen who is the claimant's own child, an adopted or step child or an illegitimate child in circumstances where the tax payer has subsequently married the mother. Also the allowance will be given if the tax payer maintains a child under the age of seventeen at her or his own expense. (The child may be over sixteen if in full-time education).

In circumstances where two or more people may be entitled to claim for the relief then the allowance is proportioned by agreement between the parties or in default of agreement, in the proportion to the time the child is resident with them in the tax year.

(d) Blind Person's Relief (ICTA 1988 s.265)

A registered blind person (or a person whose wife is registered) is entitled to a deduction, (currently £1080). Should both husband and wife be blind each will receive the deduction.

(e) Life Assurance Relief (ICTA 1988 ss.266–274 and FA 1988 s.29)

Income Tax relief can be claimed for premiums paid on a life assurance policy taken out on the life of a tax payer or his wife before 14 March 1984 if the policy satisfies certain conditions. Life assurance relief is no longer available for contracts made after that date, but it continues for policies current at that date.

1. It must be for a term of at least ten years.
2. The premiums must be evenly spread.
3. The premiums payable under the policy in any period in twelve months must not exceed twice the amount payable in any other twelve month period or one eighth of the premiums payable if the policy were to run for a specified term.

In these circumstances, the claimant is entitled to deduct from the amount which he pays by way of a premium a sum equal to 12.5% of that amount. If the total premiums paid in any year exceed £1500, then the relief is restricted to one sixth of the total income.

(Note that contributions by employees to a superannuation scheme approved by the Commissioners of the Inland Revenue are allowed as deductions in assessing salaries and wages. This is in line with the general exemption from tax afforded to savings media including pension scheme contributions and retirement and annuity premiums.)

(f) Widow's Bereavement Allowance (ICTA 1988 s.262)

When a man who is entitled to the married man's allowance dies, his widow is entitled to a bereavement allowance which is the equivalent of the difference between a married man's allowance and a single person's. From 1983–84 the allowance has been extended to cover the year after the husband's death. The relief for 1990–91 is £1720. Example 3.1, set out below, demonstrates the manner in which income may be taxed taking due account of eligible allowances.

Example 3.1

Mr Abbott, a Chartered Surveyor, lives with his wife and family at their home in Abbey Street. He pays £1800 interest on a mortgage during the financial year 1990–91. Mr Abbot's income consists of a salary from his employer of £30 000. During the year of assessment he received £300 interest from a building society savings account. His wife is working part-time as a teacher for which she receives a salary of £5000. She also receives an income of £2100 from securities in her name.

Their liability to tax for the year 1990–91 can be calculated as follows:

	Mr £	Abbott £	Mrs £	Abbott £
Earned income		30 000		5000
Unearned income: building society interest	300			
add basic rate tax at 25/75ths.*	100	400		
:dividends			2100	
add tax credit			700	2800
		30 400		7800
less mortgage interest		1 800		nil
		28 600		7800
less personal allowance		3 005		3005
married couples allowance		1 720		
Taxable balance		23 875		4795

	Mr £	Abbott £	Mrs £	Abbott £
Tax payable: basic rate £20 700 at 25%		5 175		1199
higher rate £3175 at 40%		1 270		nil
Total tax	(27%)	6 445	(23%)	1119
less credit on build. soc. interest		100		
less credit on dividends				700
Net tax payable		6 345		419

*Tax is actually deducted from bank and building society savings accounts at a 'composite rate' of tax which is usually lower than basic rate. Composite rates will be abolished from 6 April 1991 when interest will be received gross on these savings.

CORPORATION TAX

Introduction

Corporation Tax was introduced in the Finance Act of 1965 and is now chargeable under ICTA 1988 s.6. Prior to 1965 companies were liable to Income Tax plus 15% Profits Tax. Corporation Tax is payable by all companies and unincorporated associations resident in the UK on all of their profits.

Tax is charged after allowable deductions on profits wherever arising and whether remitted or not to the UK. Profits comprise income and capital gains, income being computed in accordance with Income Tax principles and capital gains in accordance with the principles of Capital Gains Tax. In computing total profits before 17 March 1987 only a fraction of a company's capital gains were taken into account so that the gains bore the same effective rate of tax as the gains accruing to an individual.

Corporation Tax is now charged both on a company's profits and its capital gains (or losses) (ICTA 1988 ss.345–347, 400 and 435). This means that the capital gains of companies are now charged to Corporation Tax at Corporation Tax rates.

Rate of tax

Rates of Corporation tax are set for financial years which end on 31 March. The rates are set in the Budget and Finance Act for the preceding year although in the Finance Act 1984 rates of Corporation

Table 3.1. Corporation tax rates

Financial Year	Tax Rate	Taxable fraction of capital gains
	Percent	
1984	45	2/3
1985	40	3/4
1986	35	6/7
1987	35	no reduction for gains made after 17 March 1987
1988 (and thereafter)	35	no reduction

Tax were fixed for the financial year 1983 and the following three years over which period substantial reductions in the rate of Corporation tax were predetermined as were the rates which apply to small companies.

Companies are assessed with reference to 'Chargeable Accounting Periods' of not more than twelve months which is a normal commercial accounting period, and the tax is generally payable by companies nine months after the end of the accounting period. Where an accounting period spans more than one financial year the amount assessable is apportioned on a time basis and the rate of tax for each financial year is applied to each separate proportion.

A company may claim the benefit of the Small Companies Rate (25% from 1 April 1988) if its profits do not exceed £200 000 (1990–91) (ICTA 1988 s.13). Where a company's profits exceed £200 000 but are less than £1000 000 the Corporation Tax charged on its income may be reduced by a fraction of the difference between £1000 000 and the company's profits. The values for this fraction in recent years has been as shown in Table 3.2.

Table 3.2. Small Companies Rate

Financial Year	Tax Rate	Relief Fraction
	Percent	
1984	38	3/80
1985	30	1/40
1986	29	3/200
1987	27	1/50
1988	25	1/40
1989	25	1/40

Marginal relief as a deduction from Gross Corporation Tax on income charged at the higher rate is computed according to the following formula: the upper marginal relief limit − profit) × Income/Profit × relief fraction.

or

$$(M - P) \times \frac{I}{P} \times f$$

Clearly the fraction will vary from year to year with changes in rates of tax and the profit limits but for the financial year 1990–91 it is 1/40. Thus giving a marginal rate of tax of 25% where the profits comprise only trading income.

Franked Investment Income

The definition of profit for the purpose of the Small Companies Rate of Corporation Tax consists of all income, capital gains and 'Franked Investment Income'.

Where distributions of company profits are made dividends are not allowable deductions when calculating profits chargeable to Corporation Tax. However, where dividends from UK resident companies form part of the income of a company then they are not required to be included in that company's profits to be chargeable to Corporation Tax since the company remitting the dividends will have already suffered Corporation Tax prior to the dividend being paid.

For this reason the dividends together with their attached taxed credits are defined as Franked Investment Income when received by the company. Where this occurs such income can be transmitted onwards as dividend to the company shareholders without the company being liable for additional tax. The shareholders who receive the Franked Investment Income then receive credit for the remitting company's tax payment to a value of 25/75 of the dividend received.

Whereas Franked Investment Income as dividends received from other UK resident companies is not chargeable to Corporation Tax, it is nevertheless included in the computation of total profits from which the rate of Corporation Tax on income is determined (See Examples 3.2 and 3.3).

Payment of Corporation Tax (ICTA 1988 ss.238–246)

Corporation Tax is payable in two ways.

1. Advance Corporation Tax (ACT) becomes due whenever a 'qualifying distribution' is paid during the actual chargeable accounting period. A qualifying distribution usually means the payment of dividends to shareholders. When a company distributes dividends to the shareholders it is assumed that the amount paid is net of basic rate of Income Tax. The company must pay this tax to the Inland Revenue and this is done by a payment of ACT. Once ACT has been

paid the amount may be set-off against Mainstream Corporation Tax (MCT). This procedure is known as 'imputation of tax'.

2. Nine months after the end of the chargeable accounting period the balance of the Corporation Tax (MCT) becomes payable.

Example 3.2

The last accounting period for a company end on 31 March 1991. Its accounts show the following income (adjusted for tax).

	£
Trading profit	1050 000
Bank interest received	1 720
Capital gains	35 000
Dividends received from other UK Companies	25 000 (franked investment income)

During the same accounting period the company paid interest on loans (debentures) of £37 000 (before deduction of tax)

The calculation of liability to Corporation Tax then proceeds as follows:

	Income	Chargeable gains	Franked investment income (Including tax deducted at source)	Profit
	£	£		£	£
Sch. D. Case I	1050 000	35 000	Net of tax	25 000	
Sch. D. Case III	1 720		credited @ 25/75ths	8 333	
Annual charge (debt interest)	37 000			
	1014 720	+ 35 000	+	33 333	= 1083 053

Gross Corporation Tax:

On income	£1014 720 @ 35% =	£355 152
On chargeable gain £	35 000 @ 35% =	£ 12 250
Corporation tax payable		£367 402

Example 3.3

A company makes up accounts for twelve months to 31 March 1991 with these results, (all figures have been adjusted for tax).

	£
Trading Profit	217 349
Bank Interest received	937
Capital gains	21 475
Franked investment income	7 471 (net of tax)

The company paid loan interest of £19 723

	Income	Chargeable gains	Franked investment income (Including tax deducted at source)		Profit
	£	£		£	£
Sch. D. Case I	217 349	21 475	Net of tax	7 471	
Sch. D. Case III	937		@ 25/75ths	2 490	
	218 286	21 475		9 961	
Less loan interest	19 723				
	198 563	+21 475	+	9 961	229 999

Gross Corporation Tax:

On income	198 563 @ 35%	= 69 497
On chargeable gains	21 475 @ 35%	= 7 516
	220 038	£ 77 013

less marginal relief

$$(1000\,000 - 229\,999) \times \frac{220\,038}{229\,999} \times \frac{1}{40} \qquad = 18\,416$$

Corporation tax payable	58 597

Close Companies (ICTA 1988 ss.414–430 and Sch.19)

Close Companies are (usually) unquoted family-owned companies. Special rules (all disadvantageous to the company and its members) apply which are intended to prevent the use of such companies for tax avoidance purposes.

A 'Close Company' is one which either is under the control of five or fewer participants and their associates, or is under the control of any number of directors and their associates.

The consequences of being a Close Company are as follows.

1. The definition of 'distribution' is extended. This means that benefits in kind given to 'participators', i.e., persons with a financial interest in the company, their 'associates' (who are not employees), i.e., ancestors, brothers, sisters, spouses descendants and business partners are treated as distributions.
2. Loans made by the company to its members are treated as distributions.

3. The company must distribute a certain level of its income. (Although this requirement has been abolished for accounting periods ending 31 March 1989. FA 1989 s.103).

4. Small companies rate does not apply to a 'close investment company' (FA 1989, s.103).

INCOME FROM LAND

Income from land is not subject to a single charge to tax. Rental income arising from the letting or ownership of property is chargeable to Income Tax or Corporation Tax. These taxes are leviable at normal rates.

Income from property might arise as profits from furnished or unfurnished lettings, receipts from rent charges and royalties, (such as minerals or tolls) profits from the cultivation of woodlands, the premiums on a grant of a lease, (although these appear to be capital payments they may be taxed as income under Schedule A), or profits from the provision of services to occupiers of land.

As a consequence of these diverse sources of income, tax is charged under a variety to schedules including A, B, and D Cases I, V and VI.

Schedule A (ICTA 1988, s.15)

Under Schedule A income from land is charged not on *rents per se* but on the annual *profits* arising by virtue of an ownership of an interest in land.

Taxation legislation does not define rent, but rent has its common meaning of payments due from a tenant to a landlord in respect of the grant of a lease.[3] For this reason premiums paid for such occupation rights may be taxed as if they were rent (ICTA 1988 ss.33–39 and Sch.2 and FA 1988 s.75).

Specifically the charge to tax on the profits arising from rents and other receipts from land are in respect of rents under leases, rent charges, ground annuals, feu duties, or other receipts arising as a benefit to an individual as a consequence of the ownership of an interest in land.

Although rent is given its ordinary meaning that meaning is extended (ICTA 1988 s.24(6)) to include any payments made by a tenant to a landlord in respect of costs incurred in repair and maintenance provided that such payments are also classified as rent payable under the lease.

In circumstances where a premium may be paid, then if the lease is granted for a period of fifty years or less, part of that premium is treated as rent (infra).

Basis of assessment

Income Tax under Schedule A is based on the amount that the tax payer is entitled to receive (ICTA 1988 s.21(1)) in the chargeable period.

The tax payer is liable therefore even although s/he may not have received the sums to which s/he is entitled. Section 41 allows the tax payer to deduct sums which s/he has not received where either non-payment was because of the default of the person by whom it was payable or if in cases of hardship, the payment of rent had been waived by the owner of the interest in land. If a deduction has been made under this provision and if rent is subsequently recovered after relief has been granted then the tax payer must give notice within six months and the income received will be treated as income for the year to which it refers.

Taxes are due on the 1 January of the year of assessment and are levied on a current year basis. Since this requires profits for the financial year to be estimated adjustments may be made to the assessment after 6 April when the actual receipts and expenses are known.

Deductions

When a calculation of income in respect of an interest in property covered by Schedule A has been made then it is permissible to make certain deductions from the gross amount in order to compute the assessable profit. (ICTA 1988 s.25). Section 31 includes some minor modifications to this basic rule including the important one (ICTA 1988 s.31(4)) that in circumstances where the amount of deductions exceeds the gross income then the tax payer may decide to deduct some only of the allowable items in that year rolling the remaining ones forward as a basis for making a future claim for losses. Clearly this provision would be of particular significance when more than one Schedule A source is involved.

Deductions may be made in respect of the following.

1. Rent, rent charges, ground annuals feu duties or other payments payable by the landlord.
2. Cost of repairs.
3. Cost of maintenance.
4. The cost of services provided by the tax payer under the terms of the lease.
5. Management expenses including the upkeep of estate offices, salaries and wages of employees engaged wholly in the management of their properties or the provision of services.[4]
6. Insurance premiums (and the cost of insurance valuations).
7. The cost of preparing accounts and arranging the collection of rent.[5]
8. The cost of advertising properties.
9. Rates and other payments of a similar nature.

10. Capital allowances on plant and machinery used in the upkeep of the property.

Special rules redeductions

1. A claim for expenditure will not be allowed as a deduction if the amount spent can be reimbursed (for example by the recovery of insurance money for repairs carried out as a consequence of storm damage) (ICTA 1988 s.31(5)).
2. The right to set off expenditure in one taxable year against another is limited. The general rule is (ICTA 1988 s.25(3)) that money expended on allowable items are only deducted from income received during the chargeable period and in respect of the particular premises.
3. Similarly it is only possible for the tax payer to claim expenditure actually made. For example the payments of annual amounts into a sinking fund for the purpose of carrying out works of repair at a future date are not allowable.[6] However, when the full cost of the work is made then that amount can be claimed in that year as a deduction.

Other properties

Section 25(7) provides that where there is allowable excess expenditure in a let property it can be set off against income from other property on the following basis.

1. That the income from a repairing lease can absorb excess expenditure from another lease let at full market value.
2. That expenditure from a tenant's repairing lease cannot be set off against another tenant's repairing lease but can be set off against income from a landlord's repairing lease or carried forward against income from the same lease (see Example 3.4).

Sporting rights

Special relief is available for sporting rights where these are let but where there may be void periods (ICTA 1988 s.28). Although rent may not be received for a particular year the landlord may nevertheless be involved in expenditure on the management of the game. However, such expenditure would not qualify as a tax deduction since the income from sporting rights is not rent for the occupation of land in the normal way and therefore is not rent payable under a lease. If the income is not rent payable under a lease then management and maintenance expenses will not be deductable.

To cover this situation the special relief permits allowable deductions against income from sporting rights to be offset against either the income from other properties (provided they are not on a tenant's

Example 3.4

Schedule A: Income from property

A landlord owns three properties. Nos. 1, 2, and 3 are all let at full rent. Nos. 1 and 2 are tenant's repairing leases while No.3 is a landlord's repairing lease. The landlord pays the rates.

Year 1

	No.1 (f.r.i.)		No.2 (f.r.i.)		No.3 (11d. reps.)	
Rent received		900		600		300
Rates	400		300		100	
Repairs (11d. only)	750		100		–	
Insurances	50	1200	50	450	50	150
Profit (loss)		(300)		150	Offset	150
Offset against No.3		150			against No.1	(150)
Carried forward		(150)		150		Nil

Net Schedule A assessment for year 1 = £150

Year 2

	No.1		No.2		No.3	
Rent received		900		600		300
Rates	400		300		100	
Repairs	250		500		–	
Insurance	50		50		50	
Expenditure b/f	150	850	–	850	–	150
Profit (loss)		50	Offset against	(250)	Offset against	150
			No.3	150	No.2	(150)
		50	c/f	(100)		Nil

Net Schedule A assessment for year 2 = £50

repairing lease and are let at a full market rent) or against income from the sporting rights in future years.

However, it must be noted that if the landlord chooses to consume these rights himself rather than lease them in any given year then the value of the rights to the extent that he enjoys that year must be set off against the allowed deductions.

Taxation of premiums on leases

Premiums received on leases exceeding fifty years are not assessable to Income Tax (but may be liable to Capital Gains Tax). Premiums received on a lease not exceeding fifty years are treated as income in the form of rent thus falling within the role of assessment of Schedule A (ICTA 1988 ss.34–39 and Sch.2, FA 1988 s.73).

Clearly the effect of this ruling could be arbitrary if it failed to make allowance for the fact that the premium is a capitalization of income attributable to the number of years comprising the term of the lease and not as income in the first year.

Hence premiums are subject to a form of tapering relief by which the income pattern of the premium is reduced by 2% for each complete year of the duration of the lease less the first year. The premium is thus charged partly to Income Tax and partly to Capital Gains Tax, the reduced amount being treated as additional rent and the amount deducted from the premium being treated as a part disposal for Capital Gains Tax purposes.[7]

Example 3.5

A lease is granted for fifteen years at a premium of £2000, the amount which will be added to the rent in the first year for assessment to Schedule A will therefore be as follows.

	£		£
Premium	2000	*OR*	2000
Assessable to			
capital gains		*less*	
less 2% × 14 × 2000	560	(15–1)/50 × 2000	560
Assessable to Income Tax	1440		1440

The premium may not be confined to a cash payment. For example if a tenant agrees to carry out certain work or to make a payment as consideration of the surrender of a lease this will be treated as a premium and its value adjusted for assessment under Schedule A above.

Where a premium is paid in respect of premises used for business purposes then the tenant may gain tax relief on that proportion of the premium which is assessed to Income Tax (£1440 in the above example). In these circumstances the tenant can treat this part of the premium as rent paid during the term of the lease and can claim a yearly allowance as follows.

$$£1440/15 = £96 \text{ per annum for fifteen years}$$

Top slicing relief
Prior to 6 April 1988 a further relief was available to individuals (ICTA 1988 Sch.2 and FA 1988 s.75). This relief known as 'top slicing' relief allowed the individual to spread the taxable amount (i.e., the £1440 and

not the whole premium of £2000) over the term of the lease with the annual amount of tax due computed as payable at the top rate for the individual tax payer.

Again taking the example above this means that the £1440 is spread over fifteen years to give £96 per year.[8] If the individual's marginal tax rate were 40% then the amount assessable for Income Tax in the first year would be

$$£96 \times 0.40 \times 15 = £576$$

The benefit to the taxpayer was that his marginal rate of tax was reduced because the annual amount was clearly considerably less than the premium received as income in the future tax year. Clearly the benefit of the relief depended absolutely on the individual's marginal rate of tax. With the removal of higher rate tax bands this relief became unnecessary.

Sale and leaseback-restricted relief (ICTA 1988 s.780)

It should be noted that in certain circumstances tax relief in respect of rental payments may be restricted. The restrictions will apply under a sale and leaseback or similar arrangement where the property is leased back to the vendor at a rent exceeding a normal commercial rent. In these circumstances or circumstances of a similar nature the deduction will be limited to the normal market expectations for a commercial rent for the term of the lease. In ascertaining what the commercial rent of a property should be then no account may be taken of future rises in rental value.

Capital allowances (see Chapter 4)

Under Schedule A allowances can be claimed on plant and machinery purchased for the repair and maintenance and management of buildings where the income charged to tax derives from rents on these premises. In the case of an industrial building (and some hotels) let to a tenant capital allowances may be claimed.

A tax payer cannot claim capital allowances under Schedule D Case VI on furniture he provides for furnished lettings (see Schedule D Case VI, supra).

Recent case law determined that expenditure borne by a tenant under a lease on items that became landlord's fixtures could not benefit from capital allowances whether claimed by the landlord or the tenant.[9]

The Finance Act of 1985 has adjusted this apparently unjust anomaly and relief by way of capital allowances, on items that become landlord's fixtures, is available on capital spent on items such as lifts, heating and air-conditioning equipment, after 11 July 1984 (FA 1985 s.59 and Sch.17).

It should be noted that there are detailed provisions in the Act in respect of which party should receive the allowance and in respect of when the fixtures are transferred between the parties.

Schedule B (ICTA 1988 s.16 and FA 1988 s.65 and Sch.6)

Woodlands

Prior to April 1988 Schedule B applied exclusively to tax charged in respect of woodlands occupied and managed on a commercial basis, that is with a view to making profit. Hence amenity woodlands are not taxed and profits from woodlands not occupied and managed on a commercial basis will fall to be charged under Schedule A.

Tax was chargeable on the assessable value which was ⅓ of the annual value of the rent which might reasonably be expected on the basis of an annual letting with the tenant responsible for rates and taxes and the landlord responsible for repairs, insurance and other expenses necessary to preserve the land in a condition to command the rack rent. The annual value envisaged a letting of the land in its natural and unimproved state (ICTA 1970 Section 531).

For the purpose of Schedule B the question of whether a tax payer was an occupier is one of fact.[10]

Under these arrangements the amount assessed was treated as unearned income. It was nevertheless notional income and tax was paid on that notional income regardless of the actual profits derived by the landowner in any given year. On the other hand where woodlands in a commercial basis were taxed under Schedule B then no claims were admitted for capital allowances. However, grants made under the Forestry Act 1967 Section 4 did not affect tax liability.

As might be imagined, this basis of Income Tax in respect of woodlands was preferential. The reason was primarily because of the national need for high production of home-grown timber, with tax reliefs being used as a form of government subsidy to supplement cash grants available to those owners of forest land who enter into dedication agreements with the Forestry Commission.

Properly income from woodlands managed on a commercial basis should fall to be charged under Schedule D. Such a change might have been made in the early sixties when other changes in the tax system occurred but it was felt that given the long-term nature of woodland investment it was inequitable to change the basis of tax that had predetermined owners' decisions to plant.

Nevertheless there was a right available to the tax payer to elect to be charged under Schedule D Case I (ICTA 1988 s. 54(1)).

As a general rule it was of advantage to the tax payer to be assessed under Schedule D Case I in the early years of the life of a forestry plantation, when management and production expenses may be relatively high and when no profit may be earned. At that stage the tax payer may be able to claim capital allowances under Schedule D Case I. Further it is generally held that Schedule D Case I is a preferable Schedule to be assessed under throughout the productive cycle of the plantation. What was of real advantage was the ability to elect to change the basis near the end of the productive cycle.

When an election was made it would be made not in respect of an individual plantation but of all woodlands managed on the same estate with the definition of any separate estates being a matter of fact.

The attraction of reverting back to Schedule B arose when arrangements whereby the transfer of the occupation of the woodlands was made by gift or sale into a trust, family company or perhaps the ownership of a relative. This device had the advantage that the election into Schedule D did not bind the new owner who would then be able to enjoy the income from the felling under Schedule B on a tax-free basis. Although such devices look extremely attractive it must be remembered that the transfer of interest in land might be liable to Inheritance Tax and or Capital Gains Tax. However, to enjoy these advantages it was not necessary to sell the property. Only a change of occupation was necessary, so that a lease could be granted. There was therefore considerable scope for effective tax planning measures.

Finally assessment under Schedule B could only be sustained where the occupier simply felled and sold timber. If the timber was brought into any form of manufacturing enterprise then it would have to be assessed under Schedule D Case I since the timber-growing enterprise had been converted into a trading operation.

The Finance Act 1988 brought Schedule B to an end. The Schedule D election has thus been curtailed but there are transitional provision until 5 April 1993. After 15 March 1988 only certain classes of tax-payer will be able to elect to be charged under Schedule D. These will include existing occupiers or persons who had entered into contracts to purchase at that date.

Schedule D, Case VI (ICTA 1988, s.18)

Furnished lettings
Schedule A is concerned only with profits arising from the ownership of land and buildings and does not cover situations where furniture and services may be included. Accordingly furnished lettings are assessed under Schedule D, Case VI.

However, if the tax payer wishes, he may separate out those payments made by way of rent which are attributable to the occupation of the land from those which are attributable to furnishings or services (or both).[12]

In these circumstances the rental payment for the land will be included in Schedule A and the payments for furniture/services will be included in Schedule D, Case VI.

There is an advantage to the taxpayer in making such an election since the rent from the land may allow a set-off to be made for losses on unfurnished properties. If such set-off were not otherwise available these losses could only be carried forward (supra).

As with Schedule A tax is chargeable not on receipts but on the annual profits. Any necessary sums spent on maintaining the property in a condition which ensures the protection of the income may be deducted when computing taxable profits. In addition it is revenue practice to give relief for wear and tear on furnishings, as distinct from normal repairs, particularly since no capital allowances are available under Schedule D, Case VI. The convention is that a figure of 10% of the rent less rates, (and other expenses normally borne by tenants), is accepted (see Example 3.6).

Example 3.6

A landlord charges a rent of £1200 p.a. for a furnished flat. The tenant is responsible for the payment of outgoings which amount to £250.

The calculation of the annual profit will then be calculated as follows.

	£	£
Rent received	£1200	£1200
less tenant's outgoings	250	
Renewals allowance @ 10%	950×0.1	95
Insurance	75	
Repairs	225	
Accounting expenses	25	325
Profit for the year to 5 April		780

Like Schedule A income included in Schedule D, Case VI is considered to be investment income. The Inland Revenue has not accepted that providing furnished letting is engagement in a trade (but see Furnished Holiday Lettings infra). More favourable tax treatment as a trading activity might be available if services are provided to residents. Although no clear distinction between letting as a trading activity (e.g., a guest house) and letting furnished accommodation is given, case law[13] suggests that even where services were provided (laundry, cleaning, etc.) this does not amount to a trade. However, if services such as meals were provided, this might be sufficient for the activity to be classified as a trade giving access to a wider range of reliefs, (see Furnished Holiday Lettings infra).

Tax is due under Schedule D, Case VI on the full amount of the net income for the tax year on 1 January. An estimated assessment is made for this purpose with any necessary adjustments being made after 5 April when the factual position is known.

Furnished holiday lettings (ICTA 1988 ss.503 and 504)

As has been seen income from property is treated as investment income and not as profits accruing from a trade or business. An exception to this general truth is the taxation of income from furnished holiday accommodation. The provisions for this were introduced in the Finance Act 1984 (s.50 and Sch.11) which introduced new rules, backdated to 6 April 1982. These rules provide for income from furnished holiday lettings to be treated as trading income although it is not actually assessed under Schedule D, Case I. In order to be treated in this way the taxable situation has to certify certain conditions.

The distinction of this income as virtually trading income is an important one since allowable losses can be offset against income to an extent not possible for investment income.

Hence, interest on loans taken out to buy or improve property for furnished holiday lettings will be allowed as a trading expense. Relief from tax via capital allowances will be permitted. Capital gains roll-over and retirement reliefs will be available (see Chapter 7), and the income will qualify as relevant earnings for retirement annuity premiums.

To qualify the accommodation must satisfy certain conditions.

1. The accommodation must be let on a commercial basis.
2. It must be available as holiday accommodation for at least 140 days in the tax year, and actually let as such for at least seventy of those days.
3. The accommodation must not (normally) be in the same occupation for a continuous period of more than thirty-one days during at least seven months of the year.
4. The accommodation must be furnished.
5. The accommodation must be available for commercial letting to the general public.

Tax is payable in respect of furnished holiday lettings in two equal instalments on 1 January and 1 July.

Artificial transactions in land (ICTA 1988 s.776)[14]

This section of the Income and Corporation Taxes Act 1988 (formerly ICTA 1970 s.488) deals with 'artificial transactions' in land and applies to land in the UK. The objective of the section is to charge to Schedule D

Case VI any profits realized on land which was purchased or developed with a view to selling at a profit.

Previously such classification of taxation liability could result in the tax payer facing charges to tax at rates of tax higher than those on capital. Since income and capital tax rates have been harmonized this is now of lesser importance.

If a property owner believes that this section may apply to a property taxation, and he prefers that the gain be assessed under Capital Gains Tax, then he can apply to the Inspector of Taxes for clearance when detailed documentary evidence is normally expected. If clearance is given, and the transaction proceeds exactly as planned and described, then any gain will be assessed in Schedule D.

NOTES: CHAPTER THREE – TAXATION OF INCOME

1. 'Schedule' here actually refers to Parts of the Act which define classes of charge (A to F inclusive) and not to Schedules at the end of the Act.
2. *Moss Empires* v. *IRC* [1937] A.C. 785.
3. Rent, n. periodical payment for use of another's property, esp. houses and lands (Chambers).
4. *Southern* v. *Aldwych Property Trust Ltd* [1940] 2 K.B. 266.
5. *IRC* v. *Wilson's Executors* (1934) 18 T.C. 465.
6. *Birmingham Corporation* v. *Barnes* [1945] A.C. 292.
7. The part not taxed as a premium could, instead of being liable to Capital Gains Tax be subject to Income Tax under Schedule D. Case 1, if the lessor deals in land (see Chapter 10).
8. The Act assumed that the £1440 was in lieu of fifteen equal payments of £96, i.e., it assumed 0% interest. However, at, say 10% return the £1440 would buy a right to fifteen equal payments of £189.32. This feature of the legislation made it even more attractive to opt for top slicing relief.
9. *Stokes* v. *Costain Property Investments Ltd* [1983] S.T.C. 406.
10. See *Russell* v. *Herd* [1983] S.T.C. 541.
11. *Scottish Heritable Trust Ltd* v. *IR* [1945] 26 T.C. 414.
12. See *Wiley* v. *Eccott* [1913] 6 T.C. 128 and *Ryall* v. *Hoare* [1923] 2 K.B. 247.
13. *Gittons* v. *Barclay* [1982] S.T.C. 390 and *Griffiths* v. *Jackson* [1987] S.T.C. 184.
14. A further discussion will be found in the section on Dealing and developing in Chapter 10.

Chapter Four

CAPITAL ALLOWANCES

INTRODUCTION

Capital expenditure cannot be written off by a trader against income when calculating income for the purpose of taxation. Equally if he depreciates his fixed assets such depreciation is not allowable as a deduction against profit. However, tax legislation allows for relief to be given on certain items of capital expenditure and these standardized depreciation allowances are known as capital allowances. Capital allowances can be set off against trading income when the expenditure occurs and/or over a number of years of the ownership of the asset. For companies capital allowances are treated as a trading expense. In the case of partnerships or individuals, however, they are created as a specific deduction.

Not all capital expenditure is eligible, only expenditure on certain specified items the most important of which, for present purpose are as follows.

1. plant and machinery;
2. industrial buildings and hotels;
3. assured tenancies;
4. agricultural land and buildings;
5. buildings in enterprise zones;
6. capital expenditure on scientific research;
7. patents and know-how;
8. mines and oil wells;
9. dredging;
10. cemeteries and crematoria.

The rules concerning capital allowances are contained primarily in the Capital Allowances Act 1968, the Income and Corporation Taxes Act 1970 and subsequent Finance Acts, particularly those of 1971, 1984 and 1986. A capital allowance bill is before Parliament in early 1990. This will be a consolidating act.

The rules provide for a system of allowances which comprise three related parts.

1. A first year allowance (or initial allowance) is permitted of a percentage of the cost of the asset.[1]
2. During the life of the asset on annual writing down allowance is permitted, (unless the first year allowance has been claimed at a maximum 100%).[2]
3. At the end of the life of the asset, (or the period of trading) a balancing allowance or charge is calculated to ensure that the tax payer has received no more and no less than his entitlement to relief by way of capital allowance. This provision is therefore wholly equitable but it does carry implications which the tax payer should consider. For example if the allowance exceeds the expenses at the time of the sale of the asset the balancing charge could be substantial and since there is no alternative but to pay the sum due in that financial year there could be a large incidence of tax in that year (possibly even greater than the savings in tax permitted by the allowances).[3]

Where regional development grants under Part I of the Industry Act 1982 are payable in respect of plant and machinery, buildings or other qualifying categories, then they will not affect the payment of capital allowances which are therefore an additional benefit. Regional development grants are payable in the various classes of development area and derelict land clearance areas.

However, other grants which may be attracted will have the effect of reducing the capital expenditure on which capital allowances will be calculated.

Between March 1984 and March 1986 there was a dramatic reduction by way of phased withdrawal of what had been generous rates of allowances on plant and machinery, industrial buildings and hotels and agricultural land and buildings (see Table 4.1). These changes were introduced as part of a programme to reform company taxation which included significant reduction in rates of Corporation Tax to compensate for the loss of allowances (see Chapter 3). The phased withdrawal has been essentially in respect of first year allowances, or initial allowances in the case of industrial, hotel and agricultural buildings. Since March 1986, therefore it has only been possible to claim for small annual allowances. There are some exceptions to these changes and these together with details of the phased withdrawal will be provided in the particular sections which follow. For the purpose of this text it will be sufficient to deal with only five of the defined categories of asset, i.e., plant and machinery, industrial buildings and hotels, buildings in enterprise zones, agricultural land and buildings and mines and oil wells.

BASIS PERIOD

For the purpose of giving relief by means of capital allowances a basis period is recognized during which the allowance may be set against profits made. For persons subject to Income Tax the basis period is the year of assessment and for those subject to Corporation Tax the basis period is the chargeable accounting period.

Table 4.1. Rates of Capital allowances

	1983–4		1984–5		1985–6		1986–7 et seq.	
	FYA* or IA	WDA†	FYA or IA	WDA	FYA or IA	WDA	FYA or IA	WDA
	%	%	%	%	%	%	%	%
Plant and machinery	100	25	75	25	50	25	0‡	25
Industrial buildings	75	4	50	4	25	4	0	4
Hotels	20	4	20	4	20	4	0	4
Small industrial workshops	100	25	—	—	—	—	—	—
Very small industrial workshops	100	25	100	25	—	—	—	—
Assured tenancies	75	4	50	4	25	4	0	4
Buildings in enterprise zones	100	25	100	25	100	25	100	25
Mines and oil wells	40	§	40	§	40	§	0	25/10
Dredging	15	4	15	4	15	4	0	4
Agricultural land and buildings	20	10	20	10	20	10	0	4

*FYA = first year allowance; IA = initial allowance.

†WDA = writing down allowance; WDAs are calculated on a straight line reduction of percentage of initial cost with the exception of plant and machinery, mines and oil wells, patents and know-how which are calculated on a reducing balance reduction of percentage of previous balance.

‡Despite the removal of FYA on plant and machinery costs some items still qualify. In these cases the disposal value is deemed to be nil so that there can never be a balancing allowance or charge on disposal. The specified items include the following.

1   Thermal insulation of trading premises.
2.   Fire safety expenditure in boarding houses and hotels.
3.   Safety measures at sports ground.

§The WDA of mines and oil wells for expenditure before 1 April 1986 is the greater of 5% of capital expenditure or output for the year divided by the estimated output and multiplied by the original capital cost. For mineral extraction within the UK the WDA is a fraction of royalty value. After 31 March 1986 a WDA of 25% is applied except in the case of expenditure on minerals where a WDA of 10% applies.

Where allowances are eligible to be set off against profits from trade taxed on a preceding year basis then the base period will be the year on which tax is to be assessed. Accordingly, a trader operating on a calendar year basis will have that year as his basis year. Hence capital expenditure incurred by the 31 December of any given year will qualify for allowance in the financial year commencing in the following April.

There are special provisions in the CAA 1968 for overlapping accounting periods; sufficient to say here that for the purpose of determining when a qualifying asset was purchased then the date will be that upon which the expenditure was incurred. For buildings this may mean that expenditure incurred over the total time of development will have to be apportioned between years of assessment.

Note, however, that for basis periods ending after 17 December 1984 the date when capital expenditure becomes allowable as a deduction against gains or profits will be the date on which the obligation to pay becomes unconditional (FA 1985 s.56).[4] In the case of buildings this could be the date upon which title passes from a seller to a vendor. Note also that capital expenditure made before a trade has actually commenced will be considered or deemed to have been incurred on the day that trading commenced.

PLANT AND MACHINERY

Capital expenditure is defined as excluding amounts which may be allowed as deductions in calculating gains or profit in respect of the employment or trade carried out by the person or company incurring the costs.[5] (CAA 1968 s.82(1)). On the other hand 'plant and machinery' is not defined in the relevant legislation and this has been left to case law. 'Machinery' appears to be well understood but the question of what is 'plant' has been the subject of frequent litigation. The matter is clearly important where buildings are involved.

It has been held that 'plant' does not include the place where the business is carried on,[6] so that the machinery housed inside a building can be clearly distinguished from the structure of building itself. Plant comprises those items with which a particular trade is carried on and it therefore excludes the place or building where the trade is practised. It can happen that there may not be a clear distinction between the shell of a building and the machinery used in it. In these circumstances a structure which fulfils the function of plant *is* plant and not building.[7]

If capital expense is incurred by way of buildings specifically provided for the purpose of installing plant or machinery then the cost of these works will qualify for allowances as if they were themselves part of the plant or machinery.[8]

It has proved less easy to distinguish between buildings and apparatus. For example, it is clear that free-standing light fittings are separate from those attached to the building, but the wiring leading to these free standing fittings will be part of the building,[9] indeed lighting will not usually be considered to be plant unless it can be demonstrated to have a specific and specialized function.[10] For this reason air-conditioning equipment may qualify as 'plant'.

Another important consideration as to the eligibility of plant and machinery for capital allowances is that the asset must belong to the claimant (FA 1971 s.45 (1) and s.46 (2)). For example, in the case of Stokes v. Costain Property Investment Ltd, lifts were installed in a building by the tenant who sought allowances on capital expenditure in respect of plant. However since the lifts immediately became the property of the landlord they therefore did not 'belong' to the tenant.[11] Since 11 July 1984, however, capital allowances have become available to tenants who install items (such as lifts and air conditioning equipment) which then become landlord's fixtures (FA 1985 s.59, Sch.17).

The distinction between 'plant' and 'buildings' has been of greater significance in the past where there were differences between the rate of allowances available (and in favour of plant and machinery, see Table 4.1). This difference is now confined to writing down allowances and in absence of first year or initial allowances the absolute monetary differences are much less than previously.

Available allowances

Allowances are available for the chargeable period during which the expenditure has been incurred regardless of whether the plant and machinery is ever actually used. If plant and machinery is delivered in one period and the payment for it, although due, is not made until a later date it is the chargeable period within which delivery takes place to which the allowances will be credited. If, however, the acquisition of plant and machinery takes place before actual trading commences then it is deemed to have been incurred upon the latter event.

First year allowances (FA 1971, s.42)
The rate of the first year allowance was 100% for a considerable period of time but it was progressively reduced from March 1984 to March 1986 (FA 1984 s.58 and Sch.12) and since then no first year allowance has been available (see Table 4.1).[12]

The detailed application of the previously obtaining rules in respect of first year allowances could be used to optimize the relief available, but these have now become ineffective with the withdrawal of this allowance.

Writing down allowance (FA 1971, s.44)

Prior to March 1986 this was an allowance that was available when the first year allowance had been waived in whole or in part or at times when the 100% allowance was not available or again where the asset did not qualify for the 100% allowance.

Capital expenditure which, prior to March 1986, had not been the subject of a first year allowance, and all capital expenditure after that date can attract a writing down allowance at the rate of 25% by the reducing balance method.

Qualifying expenditure on plant and machinery is normally included in a single 'pool' (aggregated from expenditure on all items of plant and machinery), and the writing down allowance is then applied to the value of the pool (FA 1981, s.44(2)).

Certain types of plant and machinery must be pooled separately and these are as follows.

1. Assets used partly for non-business purposes (such as private use).[13]
2. Assets which are to be disposed of within five years, at the tax payer's option.[14]
3. Motor vehicles costing over £8000.[15]
4. Ships.

Balancing allowances and charges

When plant and machinery is sold the sale price or market value is subtracted from the pool after adding the original expenses of acquisition before a writing down allowance is calculated on the net balance. Clearly it would be inequitable for any figure larger than the original cost to be deducted from the pool and if the disposal proceeds exceed the value of the pool a 'balancing charge' must be paid to the Inland Revenue. This charge may be added to the trading profit (although it is not income), or may be used to reduce the value of the capital allowances. Should a disposal result in a value less than the value of the pool then a 'balancing allowance' will be granted.[16]

Example 4.1

A trader buys and sells items of plant and machinery for use in his business over the two base period years ending on 31 December 1985 and 1986.

January 1985	Value of the pool at start of financial year	10 000
June 1985	Sales proceeds	6 000

December 1985	Purchase expenses		12 000
June 1986	Sales proceeds		3 000
December 1986	Purchase expenses		15 000

Calculation of capital allowances

Financial Year 1985–1986
(base period year to 31 December 1985)

Value of pool		b/f	10 000
Less sales proceeds			6 000
			4 000
WDA @ 25%			1 000
			3 000
Add Purchase	12 000		
FYA allowance @ 75%	9 000		
Amount not available as FYA	3 000		3 000
Balanced forward			6 000

Financial Year 1986–87
(base period year to 31 December 1986)

Value of pool		b/f	6 000
Less sales proceeds			3 000
			3 000
WDA @ 25%			750
			2 250
Add Purchase	15 000		
FYA @ 50%	7 500		
Amount not claimed as FYA	7 500		7 500
Balance forward			9 750

INDUSTRIAL BUILDINGS (CAA 1968, s.1)

A person or company can gain the benefit of allowances on capital expenditure on the construction of an industrial building (or industrial structure) provided that it is intended to be occupied for the purpose of carrying out trading operations. Allowances are confined to the expenses of construction and although this can include site preparation works (CAA 1968, s.9) this specifically excludes the cost of the land on which the building is constructed (CAA 1968, s.17(1)).

Capital expended upon an existing building by way of improvement or refurbishment can be eligible for allowances but when this occurs the expenditure will be treated as if in a separate building.

A question which has to be addressed is 'what is "an industrial building or structure" for the purpose of the Capital Allowances Act?' The intention of the statutory definition is to identify industrial buildings where productive processes will be undertaken, as against service or distributive activities. Nevertheless those activities which qualify cover a surprisingly wide range, particularly where the fact to be established is whether the building is to be used for the purposes of a qualifying trade rather than whether the building is to be used exclusively for a particular qualifying purpose. A building may qualify even if it is only used to store goods provided that storing goods is in fact the nature of the trade.[17] The repair and maintenance of goods is a qualifying activity, provided that these goods are themselves used in a qualifying trade.[18] Activities as diverse as ploughing or cultivating land, where the trader does not himself occupy the land, and bridging and tunnelling will qualify.

However certain activities are specifically excluded and these include the use of a building as a dwelling house, hotel, wholesale warehouse, shop or office.[19,20]

Given changes in the character of industrial activity the exclusion of 'offices' may give rise to difficulty where a qualifying activity requires a significant proportion of ancillary office accommodation in the same building. This difficulty may appear to be compounded by the publication of a new Use Classes Order and the establishment of Class B1 permitting a variety of business uses. Case law suggests that the facts of the situation rather than arbitrary descriptions of accommodation will provide the basis of any test of eligibility for allowances.[21] In any event the Capital Allowances Act provides that where a non-qualifying activity occupies less than a quarter of the building in question then the whole cost will be allowed.[22]

Available allowances

Initial allowances, where available, may be claimed by the person who bears the cost of the building. Writing down and balancing allowances may be claimed by the owner of the 'relevant interest', that is the interest to which the original owner was entitled when he carried out the initial capital expenditure.[23] Hence, if an owner of the entire interest in the property carries out the development of a building and then leases it for occupation by a tenant then he will be able to continue to set-off the benefit of capital allowances against his income (the rent received), provided always that the tenant is using the building for a qualifying purpose.[24] Equally, a lessee who spends capital improving a building can claim the benefit of allowances (if eligible), while his landlord may not.

Despite the apparent simplicity of the foregoing a number of detailed situations regarding eligibility for allowances arise where leases and other arrangements are entered into between parties in respect of the expenditure of capital and occupation of buildings. Two of the more important ones are dealt with here.

Firstly, the increasing involvement of financial institutions in property since the passing of the Capital Allowances Act exposed an anomaly which has since been rectified. If an insurance company or pension fund wished to construct an industrial building as part of its property investment portfolio then they would be the owners of the relevant interest and the only possible claimants for relief under the Act. But a financial institution such as a pension fund will pay no tax on the investment income accruing from this property. Since the relief from taxation is only available by setting-off capital allowances against tax payable on income the fund would be incapable of receiving any benefit. The Finance Act 1978 (s.32) corrected this situation by overriding the 'relevant interest' rule. It arranged that where a long lease is granted,[25] the lessee can claim the allowances even although he did not incur the expenditure, provided that both the lessor and lessee agree to this arrangement and that claims for the redesignation of the relevant interest are made within two years of the commencement of the lease.

Secondly, there has been an increasing desire by Government to assist the starting up and development of small businesses. At the same time it has been appreciated that it is unrealistic to expect new businesses to enter into formal contractual leases with considerable obligations. Many small businesses prefer to occupy industrial space on the basis of a licence from the owner. Since a licence is not an interest in property they cannot then claim industrial building allowances. The Finance Act 1982 (s.74) has therefore provided that where an owner or a lessee grants a licence then the interest will be treated as subject to a lease and it will be possible for a claim for allowances to be made.

Initial allowance
The initial allowance for industrial buildings prior to 1984 was 100%. This allowance was phased out so that since 1986 no initial allowance can be claimed on industrial buildings other than those in enterprise zones (see infra). As part of a measure to encourage the start-up of small businesses over a short number of years a rate of 100% initial allowance was afforded for 'small' (<2500 sq. ft) and 'very small' (<1250 sq. ft) industrial workshops. These allowance have also been withdrawn.

Where initial allowances remain available they may be claimed either in whole or in part. In the latter case the remaining expenditure is written off on a straight line basis.

Writing down allowance

The annual writing down allowance is at a flat rate of 4% of the construction cost.[26] This allowance may be claimed in the first year in addition to the initial allowance where this is claimable.[27] Although writing down allowances may be claimable as from the first year they are not available until a letting takes place if the building has been purchased for investment rather than occupation. If a letting proves to be to a person or company engaged in a non-qualifying activity any initial allowance previously given will require to be paid back, (by assessment under Schedule D, Case VI).

If the industrial building on which a writing down allowance is claimed is second-hand then the allowance is calculated with reference to the qualifying cost, which in no circumstances can exceed the original cost of construction.

Balancing allowances and charges

If a building is sold, ceased to be used for a qualifying activity or demolished within twenty-five years of first being used, then a balancing allowance or charge requires to be assessed.[28] The allowance or charge will depend upon the 'residue of expenditure' which is simply the difference between the original cost of the building and the allowances made to that date – whether initial or writing down.[29] The residue of expenditure is then compared with the amount of consideration received for the industrial building which may be by sale, insurance (if destroyed) or scrap value (if demolished). Where the residue of expenditure exceeds the monies received then the difference is a balancing allowance. Where the residue of expenditure is less than the sum of allowances already received then a balancing charge arises.[30]

Example 4.2

(A) An industrial building is erected for the sum of £250 000. There is no initial allowance but a writing down allowance of £10 000 p.a. is claimed from the first year. After ten years the building is sold for £100 000 (the land element being treated separately.)[31]

	£
Original cost of building	250 000
Allowances received	100 000
Residue of expenditure	150 000
Sale proceeds	100 000
Balancing allowance	50 000

(B) If the building had continued to be owned and used for qualifying purpose for twenty years, destroyed by fire, and the owner compensated with the sum of £75 000 then the outcome would be as follows.

	£
Original cost of building	250 000
Allowances received	200 000
Residue of expenditure	50 000
Insurance monies received	75 000
Balancing charge	25 000

Where a building is sold there will be a balancing adjustment made in respect of the vendor but there will be a possible claim for relief by the purchaser provided that he uses the building for a qualifying purpose. The purchaser cannot claim an initial allowance, but he is entitled to a writing down allowance based on the residue of expenditure. This sum is calculated by adding any balancing charge or deducting any balancing allowance from the remaining unrelieved expenditure of the vendor. The sum is then spread in equal annual amounts over the remaining 'tax life' of the building.[32]

HOTELS

The Capital Allowance Act of 1968 did not include hotels in the elaborate definition of an industrial building or structure (CAA 1968, s.7(1)) since hotels were not included within the list of trades permitted. Special provisions have therefore been incorporated in the Finance Act 1978 (s.38 and Sch.6) which permitted an initial allowance of 20% and an annual writing down allowance of 4% on a straight line basis. Since 1986 the initial allowance has been removed and, in line with the provisions for industrial buildings tax payers who possess a relevant interest are now only entitled to the writing down allowance which remains at 4%.

In order to be a 'qualifying hotel' within the meaning of the legislation (FA 1978 s.38(3); FA 1985 s.66) the hotel (or extension to an existing hotel) must be a permanent building which is open at least four months of the year between April and October and there must be a minimum of ten bedrooms available for commercial use. Any costs for providing living accommodation for the owner will not be allowable. In addition to providing letting bedrooms the hotel is also required to offer paying guests services such as the provision of meals, cleaning rooms, making beds etc.

BUILDINGS IN ENTERPRISE ZONES

In an area designated as an enterprise zone by the Secretary of State expenditure on industrial and commercial buildings qualifies for 100% initial allowance provided that expenditure on the building is incurred within ten years of the land being designated as an enterprise zone.[33] The allowance can be claimed in respect of any buildings, other than dwelling houses, provided that these are to be used for trading or professional purposes.

The owner of the relevant interest may disclaim the whole or part of initial allowance when the balance remaining will be written off (on a straight line basis) at 25% of the cost each year until it is written off in full. Balancing allowances or charges will apply on disposal of the building within twenty-five years.

Reference to Example 4.3 will illustrate the scale of benefit involved. The ability to secure high receipts in the early years of an investment causes considerable advantage to accrue to the investor who can claim 100% initial allowances. In the particular (simplified) case the investor stands to make a target rate of return 63% higher than the rate at which he judges investment would be profitable. Alternatively he could recoup a capital gain of £363 265 on the investment.

The example presupposes that the rental value of the premises in both locations is the same and this may not be the case where initial allowances are given in relatively higher rented areas. However, the comparative investment advantage (*all other things being equal*) would be considerable from the point of view of tax efficiency in the use of capital.[34]

Example 4.3

An investor is considering the development of industrial property in two alternative locations. Building costs and rents are identical. One site is an enterprise zone. The expenses incurred are estimated to be £1 500 000 inclusive and net rents (received annually in arrears) are anticipated to be £200 000 p.a. The building is expected to have a residual value of £100 000 after twenty-five years. The investor can set capital allowances off against taxable profits. Tax is paid one year in arrears. He has sufficient profits per annum to be able to take full advantage of a 100% initial allowance, and he has a marginal tax rate of 50%. His criterion rate of return is 7.5% p.a.

(a) Investment appraisal of project outside enterprise zone

Annual and residual cash flows

1	2	3	4	5	6
		Income Tax	Capital		Cash flow
Year	Rental	@ 50%	allowances	Tax saved	(2 − 3 + 5)
	£	£	£	£	£
0	—	—	60 000	—	(1 500 000)*
1	200 000	—	60 000	30 000	230 000
2	200 000	(100 000)	60 000	30 000	130 000
24	200 000	(100 000)	60,000	30 000	130 000
25	200 000	(100 000)	(100 000)†	30 000	130 000
26)	—	(100 000)	—	(50 000)	(150 000)
26)	—	—	—	—	100 000‡

Residue of expenditure	0	NPV = £34 500	
Proceeds	100 000	IRR = 7.78%	
Balancing charge	100 000		

(b) Investment appraisal of project inside enterprise zone

Annual and residual cash flows

1	2	3	4	5	6
		Income Tax	Capital		Cash flow
Year	Rental	@ 50%	allowances	Tax saved	(2 − 3 + 5)
	£	£	£	£	£
0	—	—	1 500 000	—	(1 500 000)*
1	200 000	—	—	750 000	950 000
2	200 000	(100 000)	—	—	100 000
24	200 000	(100 000)	—	—	100 000
25	200 000	(100 000)	(100 000)†	—	100 000
26)	—	(100 000)	—	(50 000)	(150 000)
26)	—	—	—	—	100 000‡

Residue of expenditure	0	NPV = £397 765	
Proceeds	100 000	IRR = 12.70%	
Balancing charge	100 000		

* Original capital cost of building.
† Balancing charge.
‡ Residual value of building.

DWELLING HOUSES LET ON ASSURED TENANCIES

A new system of capital allowances was introduced for an experimental period of five years in respect of qualifying dwelling houses which were constructed with a view to being offered for let on 'assured tenancies' within the meaning of the Housing Act 1980 s.56. The Finance Act 1982 (Sch.12, para. 8(1)) provided that where expenditure was incurred after 9 March 1982 and before 1 April 1987 an initial allowance of 75% would be permissible together with a writing down allowance of 4%.

In the interim the rate of initial allowances was scaled down and the present position is that expenditure on the development of dwelling houses let on assured tenancies under Part I of the Housing Act 1988 (HA 1988 s.95) is limited to an annual writing down allowance of 4%.

If capital allowances are to be claimed under these provisions the landlord must be a company (FA 1982, Sch.12 para.3(3)(a) and F(2)A 1983, s.6).

A dwelling house will not qualify if any of the following apply.

1. The landlord is a housing association or self-build society (FA 1982, Sch. 12 para. 3(3) (b)).
2. The landlord and tenant are connected persons.
3. The landlord is a close company and the tenant is a participator.
4. Any arrangement exists between landlords of different dwelling houses under which one landlord takes as a tenant a person who if they were to be a tenant of the other landlord, the dwelling would not be a qualifying dwelling by virtue of 2 or 3 above (FA 1982, Sch.12 para.3(3)).

A balancing adjustment will arise if, within a period of twenty-five years after a dwelling house is first used, the relevant interest therein is sold, or if it is a leasehold interest, it comes to an end otherwise than on the persons entitled thereto acquiring a reversionary interest, or it is demolished or destroyed or ceases altogether to be used (FA 1982, Sch.12 paras.4–6).

Finally when calculating any loss on the disposal of the building for capital gains tax purposes any expenditure for which capital allowances have been given is not deductible.

MINES AND OIL WELLS

Capital allowances are available for industrial buildings, machinery and plant where these are confined to the searching for discovering, testing and exploiting sources of mineral deposits of a wasting nature and where the buildings and machinery so constructed will be of negligible value when the mineral is exhausted (CAA 1968, ss.7, 51 and 52).

Expenditure on works likely to be of little value on completion of the operation qualified for an initial allowance of 40% until 1986. This is now removed and only a writing down allowance is now available.

The writing down allowance in the case of mines and oil wells was calculated by reference to a fraction which was the larger of

$$\frac{1}{20} \quad or \quad \frac{\text{the output during the period}}{\text{output during the period and estimated output of resource until depleted.}}$$

In the case of mineral depletion specifically within the United Kingdom (as against overseas) the method of calculating the allowance was on a completely different basis. Here an arbitrary fraction of hypothetical 'royalty value' was derived. The fraction varied in accordance with the time elapsed since the resource was first developed, hence the following formula.

50% of royalty value up to ten years after the costs were incurred.
25% of royalty value up to twenty years after the costs were incurred.
10% of royalty value beyond twenty years.

The writing down allowance, therefore, was a fraction of the 'royalty value' of the output (during the period) less the royalties actually paid. From 1 April 1986 it will be a straight 25% or 10% for minerals.

AGRICULTURE AND FORESTRY

An owner or tenant of agricultural or forestry land may claim allowances for expenditure on capital improvements to the land itself (e.g., drainage) or on the construction of farmhouses, fencing or other appropriate works (CAA 1968 s.68(1) and ICTA 1988 s.33). Although allowances may be claimed for improvements to land, no allowances can be claimed for the purchase of land (as with industrial and other buildings and works.[35] The allowance is not available under Schedule B.[36]

Before 1 April 1986 an initial allowance of up to 20% of expenditure was permitted, but since that date the initial allowance has been withdrawn. Similarly a writing down allowance of 10% per annum has previously been permitted but this is now reduced to 4% per annum, until the total expenditure has been allowed.

From being in a different, and somewhat privileged position relative to allowances for other buildings (the whole benefit of allowances could be recouped within eight, or a maximum of ten years), the allowances of agricultural and forestry buildings and works have now been brought into line with provisions for most other assets. However, one important difference remains, and that is that on sale of the assets there is no balancing adjustment.[37] The new owner of the relevant interest will receive the benefit of the outstanding allowances provided always that his is a qualifying purpose (FA 1986 Sch. 15 para. 4).

To qualify for the allowances in the first instance the costs must have been expended on buildings or works specifically related to husbandry or forestry.[38] If the use of a building is only partly for agriculture or forestry then an apportionment of the capital expenditure may be made to determine the amount which will attract allowances (CAA 1968, s.68(2)).

In the case of a farmhouse (i.e., a house occupied by the person running the farm) not more than one third of expenditure will qualify for relief. The fraction may be even smaller if it is considered that the qualities and characteristics of the accommodation are out of proportion to the size and type of farm.[39]

For agricultural and forestry allowances a basis period of 1 April to 31 March of the year prior to the assessment is normally taken, although where the Inspector of Taxes agrees the owner's accounting period can be used, where it is different (CAA 1968, s.68(6)).

Example 4.4

Smith constructs a barn at a cost of £50 000 completing the building in September 1987. He sells the building to Jones for £60 000 in 31 May 1991.

	Year	£	Balancing Charge	£
Smith	1987/88 WDA @ 4%	2000	Orig. cost of bldg.	50 000
	1988/89 WDA @ 4%	2000	Allowance received	6 333.33
	1990/91 WDA @ 4%	2000	Residue of expenditure	43 666.67
(year of sale)	1931/1992 WDA @ 4% × ⅙ (balancing charge which is not required to be paid by Smith)	333.33	Sale proceeds	60 000
				6 333.33*
Jones	1992/1993 WDA @ 4% × ⅚	1666.67		
	1993/1994 WDA @ 4%	2000		
	1994/1995 WDA @ 4%	2000		
	1995/1996 WDA @ 4%	2000		
	(until total allowances = £50 000)			

*The balancing charge cannot exceed the amount of the allowances made (CAA 1968 s.6).

NOTES: CHAPTER FOUR – CAPITAL ALLOWANCES

1. The allowances are given automatically to companies but not to individuals who must therefore claim. On occasions it may be advantageous to the tax payer *not* to claim them (or disclaim them if a company).
2. Initial allowance at 100% can only now be claimed for capital expenditure on commercial or industrial buildings in enterprise zones (FA 1980 s.74 and Sch.13).
3. Note that the balancing charge is limited to the amount which had initially been allowed for capital allowances, any surplus over that figure realized on the sale of the asset will be subject to capital gains tax. A balancing charge is not 'income', see *IRC* v. *Wood Bros (Birkenhead) Ltd* [1959] A.C. 487.
4. This brings the rules for capital allowance into conformity with accepted accounting practices.
5. *Rose and Co (Wallpaper and Paints)* v *Campbell* (1968) 1 All E.R. 405.
6. *Jarrold* v. *John Good and Sons Ltd* (1969) 1 All E.R. 141.
7. *IRC* v. *Barclay, Curle and Co Ltd* [1969] 1 All E.R. 732.
8. *IRV* v. *Barclay, Curle and Co Ltd (supra)*, and CAA 1968 s.45.
9. *J Lyons and Co Ltd* v. *A–G* [1944] 1 All E.R. 477.
10. *Cole Bros Ltd* v. *Phillips* [1980] S.T.C. 518.
11. [1983] S.T.C. 406.
12. But note that payment made after 13 March 1984 and before 1 April 1987 and in performance of a contract entered into before the earlier date continued to qualify for the 100% allowance.
13. FA 1971 Sch.8, para.5.
14. FA 1985, s.57 and Sch.15.
15. FA 1971, Sch.9(10); F(2)A 1979 s.14; FA 1980 s.64; FA 1984 s.61.
16. In previous years a balancing charge could sometimes be avoided by not claiming all of an FYA and then adding to the pool the remaining expenditure on which FYA was not claimed. The opportunities to exercise this option are now limited to qualifying buildings in enterprise zones.
17. See *Crusabridge Investments Ltd* v. *Casings International Ltd* [1979] 54 T.C. 246 and *Saxone Lilley and Skinner (Holdings) Ltd* v. *IRC* [1967] 1 All E.R. 756.
18. FA 1982 s.75.
19. CAA 1968 s.7(3).
20. There are special provisions for hotels (infra).
21. *IRC* v. *Lambhill Ironworks Ltd* (1950) 31 T.C. 393.
22. CAA 1968 s.7(4) FA 1983 s.30.
23. CAA 1968 ss.2(1)(c), 3(1)(2) and 11.
24. CAA 1968 ss.6, 71 and 74.
25. That is, a lease of over fifty years duration (ICTA 1970, s.84).
26. But 25% for expenditure on small industrial workshops before 27 March 1985 and for industrial buildings in enterprise zones. On buildings erected prior to November 1962 a WDA of 2% may be given.
27. Hence the designation of 'initial allowance' rather than that of 'first year allowance' for plant and machinery where WDAs accrue after the first year and on a reducing balance basis.
28. CAA 1968, s.3(1).
29. CAA 1968, s.4.
30. Subject only to the rule that the balancing charge may not exceed the sum of the allowances made (CAA 1968, s.6).

31. Since land does not qualify for capital allowances it makes more sense from the tax point of view to develop qualifying properties on leasehold land unless substantial gains are anticipated in land value.
32. The 'tax life' of a building will normally be twenty-five years where WDAs at 4% are claimed. Buildings used before November 1962 will have a 'tax life' of fifty years and buildings where an initial allowance is available and wholly or partly used, will have a 'tax life' of less than twenty-five years.
33. FA 1980 s.72. The significance of the ten year period is that it is the purpose of the enterprise zone to 'pump-prime' development in the short term.
34. When high initial allowances were available in areas other than enterprise zones the effect on land values and even the value of property company shares as a result were observed: see Stapleton T., 'Capital allowances – the neglected factor'. *Estates Gazette*, Vol. 266 pp. 499 and 501.
35. FA 1986 Sch.15 para.11.
36. FA 1986 Sch.15 para.8.
37. This is a valuable form of relief since it means that any balancing charge which might be due is not paid by the owner on disposal.
38. Note that allowances can be claimed for buildings erected for the purpose of intensive rearing. Note also that if a building does not qualify for agricultural or forestry relief it may attract industrial building allowance if the use is one which qualifies (CAA 1968 s.7(1)).
39. *Lindsay* v. *IRC* [1952] 34 T.C. 289.

Chapter Five

TAXES ON EXPENDITURE

INTRODUCTION

In recent years there has been renewed discussion on the most equitable and efficient tax system. Central to this issue is the identification of the basis on which individuals can be most accurately assessed and levied on their 'ability to pay' taxes. Presently the taxation system in the United Kingdom recoups taxation under the three principle means of identifying 'ability to pay', viz. income (Income and Corporation Taxes), wealth (taxes on capital), and consumption (taxes on expenditure).

The present Government has displayed a consistent desire in recent years to transfer the burden of taxation from income and wealth to consumption, in the belief that this will meet requirements of equity while avoiding some of the inhibiting features of capital and income taxation. Taxes on consumption are progressive since patently a wealthy person is more likely to spend, and therefore to incur tax, than a person with limited means. Whether the rich person is making the same sacrifice (in paying tax on expenditure) as the poorer person remains arguable. However, it is equally debatable whether the rich person is making the same sacrifice (in paying tax on income) as the poorer person.

Given such uncertainties, (never mind the uncertainties related to who benefits and to what extent from government services provided from tax revenues), it is probably true to say that no water-tight case can be made absolutely for one system of taxation as against another. The decision as to whether a system of taxation on one basis or another is selected is therefore prone to subjective views of how alternative systems may work in practice and also to political beliefs as to the overall benefits to society of one system over another.

It is not the objective of this text to explore the theoretical arguments which have been made in favour of or against particular tax systems. Sufficient here to observe firstly, that the overriding justification for taxing an individual on his consumption is that it is more equitable to tax him on the value of the goods that he takes out of society (i.e., the goods and services that he consumes) then on the value of his inputs to societal well-being (i.e., earnings on his services, interest on capital etc.) Secondly, that while there may be imperfections in a system of taxation based on expenditure, there are imperfections in other systems, but any logically sound system is preferable to the unstructured tax system which has obtained in the United Kingdom. Thirdly, the present government is already reducing capital taxation and income taxation and is actively seeking alternative consumption taxes with a view towards moving to a system of expenditure taxes.

The two taxes which are discussed in this chapter are quite different in nature but are distinguished as expenditure taxes. The first, Value Added Tax (VAT) is a tax on consumption, the second, Stamp Duty is a tax on legal instruments.

The reader should also be aware that there are a large variety of taxes on expenditure other than VAT and Stamp Duty. These include excise duties (on tobacco, alcoholic drink, hydrocarbon oils, betting and gaming and matches and mechanical lights), vehicle licencing tax and miscellaneous licences such as television licences.

VALUE ADDED TAX

Introduction

Value Added Tax is not an instrument conceived by the United Kingdom government as part of its complicated taxation system. It has been introduced to this country from the European Community under EC Directives, the objective of which are to seek harmonization of various turnover taxes.[1] The original directive dates from 1969 and the tax was introduced into the United Kingdom on 1 January 1973 by means of the Finance Act 1972. Subsequent amendments to the legislative provisions were made in 1971 specifically to take into account the requirements of the EC Sixth Directive and the main legislation is now embodied in a consolidating instrument, the Value Added Tax Act 1983.[2] However many of the regulations in respect of VAT are contained in statutory instruments which are issued under the authority of Section 45 of the VAT Act 1983.

The UK had resisted EC pressure to harmonize VAT taxation between member states and was in breach of the EC Sixth Directive of May 1977. The EC pursued the matter in the European Court of Justice where a ruling was made in favour of the EC on 21 June 1988. The Court ruled that zero rating was lawful in respect of private housing accepting that this was a facet of social provision. But it ruled that zero rating was not lawful in respect of the construction of buildings for industrial and commercial use. The Government implemented the necessary changes in the Finance Act 1989.

THE NATURE OF THE TAX

The tax is levied on consumers, specifically on the goods or services supplied by a taxable person in the course of business. It is a multi-staged tax. Valued Added Tax is added at each stage in the supply of goods and services. At each stage the tax is exactly proportional to the price of goods and services. VAT regulations require therefore that when a trader sells goods or provides services VAT must be added to his price. This is known as 'output' tax. When the same trader buys goods or services he himself must pay to his supplier tax over and above the price of these services. This is known as 'input' tax. The difference between output tax and input tax is the tax on the value that the trader has added to the goods or services, and this is the amount of tax that he as an individual will pay to the government. Example 5.1 illustrates the process of adding value and the incidence of tax at each stage. It shows how the tax collected at each stage totals to the same amount of tax borne by the final consumer.

Example 5.1

A timber merchant supplies dressed timber to a window manufacturer who uses the timber to make window frames. The window frames are sold to a wholesaler who sells them to a builder who intends to use them in connection with his repair and construction business.

The sequence of events is therefore as follows.

	Purchase	Sale	Input	*Tax at 15%* Output	*Tax* Paid
	£	£	£	£	£
Timber merchant		100		15	15
Manufacturer	100	150	15	22.50	7.50
Wholesaler	150	200	22.50	30	7.50
Builder	200	250	30	37.50	7.50
Client	250		37.50		37.50

Hence, the final consumer of the product bears the full impact of the tax since he cannot off-set it against any one else. At each stage of the process the trader pays tax to the government to the extent that he has added value, but this amount is recouped. Only the final consumer is unable to recoup.

Administration of VAT

VAT is administered by the Board of Commissioners of Customs and Excise.[3] The Board is therefore responsible for managing the collection and enforcement machinery for VAT, and it is also responsible for formulating policy decisions in respect of the determination and collection of the tax.

Broadly, the Board of Commissioners arrange for administration and collection of the tax through two separate directorates.

1. *Administration Directorate.* This office, based in London, is responsible for setting policy for the VAT system in respect of both general and detailed matters.
2. *Collection Directorate.* This office, based in Southend, but with local offices throughout the country, is responsible for registering tax payers, delivering VAT returns and collecting tax due or making repayments. Local offices are also responsible for examining the books of registered tax payers at regular intervals, (usually within a two to three year period).

Taxable persons

A taxable person is one who makes taxable supplies while he is, or is required to be, registered for VAT.[4] A person is required to be registered for VAT if the value of her/his taxable supplies in a period of one year exceeds £25 400.[5]

When this condition is met a person becomes liable to registration and it is then her/his responsibility to pay VAT on her/his value added whether or not she/he is able to collect VAT from those to whom she/he has made taxable supplies, but some relief for bad debt will be available from April 1991.

Similarly there are requirements to be met and procedures for deregistration should a taxable person's value of taxable supplies fall below regulated limits.[6]

Registration

It is not necessary to meet the statutory requirements in respect of value of taxable supplies in order to register. A non-taxable person whose supplies do not exceed the limits may apply to be registered for VAT purposes so that s/he may reclaim the input tax that she/he suffers. The Commissioners have discretion in such matters and may specify a minimum time period for registration (say two years). When s/he is registered s/he then possesses a VAT number which makes it possible for her/him to collect VAT from customers.

An example of a person who may wish to seek voluntary registration would be one who only makes supplies to other registered persons. Hence, if a Chartered Surveyor in private practice only provided service to persons or firms who are registered, then it would be an advantage to him to register in order to be able to offset the input tax that he suffers. However, he will also then be liable for keeping proper VAT accounts and making VAT returns and this will undoubtedly mean a degree of trouble and expense.

Businesses liable to VAT

The term business is intended to include any trade, profession, or vocation. However its application is wider than the preceding form of words would suggest. VAT can apply to *any* establishment located in the United Kingdom from which goods and services are supplied. It therefore includes admission to premises and clubs, for example. It is not even necessary that the activity be profitable since the tax is only concerned with turnover and not with profits.[7]

The concept of business is clearly different from that subject to liability to tax under Schedule D Cases I and II.[8]

Goods and services supplied

The concept of supply is central to the understanding and operation of VAT although the concept is very widely drawn, particularly with respect to the supply of services. On the one hand the Act defines supply of services as anything which is not a supply of goods but which is done for a consideration (VATA 1983 s.3(2)(b)).

It would appear therefore that any transaction involving the transfer of ownership of goods or where services are supplied for consideration is 'supply' and therefore may be liable to tax.

Detailed consideration is given below to supplies of interests in land. In the meantime examples of supplies of goods and services are as follows.

1. Supply of goods.
   (a) Outright sale of a good.
   (b) Sale by hire purchase.

   (c) Conditional sales.

   (d) 'Sale or return' agreement.

   (e) Part exchange.

   (f) Supply of power, heat, refrigeration or ventilation.

   (g) Grant of a major interest in land.[9]

   (h) Production of goods by processing other person's materials.

2. Supplies of service.

   (a) Professional services.

   (b) Agent's services.

   (c) Transfer of undivided share of property of goods (e.g., by hire or rental).

   (d) Grant of a right over goods (e.g., copyright or licence).

3. Certain transactions are deemed to be supplies and these include the following.

   (a) Transfer of assets which cease to form part of a business but which remain the property of the supplier (VATA 1983 Sch.2 paras.5(2) and 4(1)).

   (b) Gifts made in the furtherance of business (VATA 1983 Sch.2 para.5).

   (c) Goods forming assets in a business which is conducted by a taxable person and which are supplied by him after he ceases to be a taxable person if the value of the supplied goods is greater than £250 (VATA 1983 Sch.2 para.7).

4. Transactions deemed not to be supplies include the following.

   (a) The transfer of a business as a going concern (VATA 1983 Sch.7 para.4)[10]

   (b) The repossession of goods under a finance agreement.

   (c) The loss or destruction of goods.

Time of supply

The time of supply (or the tax point) occurs when the goods or services are supplied. Normally this will be the date of issue of the tax invoice. There are detailed provisions concerning the time of the supply of goods but, again normally, if an invoice is not issued within fourteen days of the dispatch of goods then the tax point will be the day on which the goods were dispatched (VATA 1983 s.5(1)).

In the case of services the matter of deciding the time of supply is simpler. A supply of a service is treated as supplied when performed. The only substantial difference is where services are supplied on a continuous basis, as say, with the provision of property management services by a Chartered Surveyor. Here the tax point will be the date that payment is received or the issue of a tax invoice (if earlier) (VAT (General) Regulations SI 1980, No.1536 Reg.21).

The value of a supply

Where a supply is made in return for a consideration in money the value of the supply will then be that amount which taken with the tax chargeable amounts to the consideration (VATA 1983 s.10(2)). From this definition of value it follows that if there is no reference to VAT in the agreement for sale, then the supplier must pay the tax out of the consideration which he receives.

There is no requirement that the consideration should be adequate, but the Act requires that where the supply of goods is not wholly for consideration in money then the value of the supply is deemed to be its 'open market value'. (VATA 1983 s.10(2)).[11]

The Act provides special rules of valuation for particular cases,[12] and it also provides various anti-avoidance provisions which require open market value to be applied where the supplier and recipient are connected persons.[13]

Rate of tax

There are two rates of tax. The standard rate of VAT is 15%.[14] The lower rate is 0%. The standard rate is applied to the supply of all goods and services by a taxable person other than those which are specifically lower rated or exempted. The reason for having a lower rate of 0% is to avoid certain goods and services which are socially necessary or desirable being subject to the tax but permitting the taxable person to recover input tax. As will be seen below certain categories of goods and services are also exempt from VAT, but in these circumstances the person who makes supplies is not registered. Since he cannot recover any input tax charged to him, in effect he has to carry that tax as a cost to his business.

The categories of supplies which are eligible for zero-rating are (briefly) as follows.[15]

Group 1: Food for human consumption and feeding stuffs for animals which produce food for human consumption. There are certain specific exemptions from this Group including the following.
(a) Alcoholic beverages (which are subject to excise duty).
(b) Chocolate or cakes, biscuits or sweets covered in whole or in part with chocolate.
(c) Food supplied from a catering establishment together with hot food sold for consumption of the premises.
Group 2: Sewerage and water.
Group 3: Books including newspapers, magazines and maps. (But note

that architectural drawings or services the end result of which is a report are not eligible for zero-rating).[16]

Group 4: Talking books and radios for the blind and handicapped.

Group 5: Newspaper advertisements were zero rated prior to April 1985 but since then they have been standard rated.

Group 6: News services. This group was repealed with effect from 1 April 1989.

Group 7: Fuel and power, i.e., coal, coke, gas, electricity and fuel oil. (Petrol and diesel fuel used in road vehicles are standard rated.)

Group 8: The grant of a major interest in a building or site (or any part of them) is zero-rated, provided that:

  (a) the building is designed as one or more dwellings or is used for residential or charitable purposes; and

  (b) the supply is made for the person constructing the building.

Services supplied in the course of constructing buildings meeting the above conditions are also zero-rated as is civil engineering work carried out in connection with a permanent site for residential caravans.

Buildings and services not so qualifying may be zero-rated under transitional arrangements.

Construction services provided by an architect, surveyor, consultant or supervisor are standard- not zero-rated.

Group 8a: The grant of a major interest in a reconstructed listed building or scheduled monument is zero-rated provided that the building is to be used for residential or charitable purposes following reconstruction and alteration and the supply is made by the person who carried out the work.

Group 9: International services performed outside the UK or supplied to a person who belongs outside the UK. Specific cases include:

  (a) services relating to land outside the UK;

  (b) the valuation of goods outside the UK if the services are performed outside the UK.

Group 10: Public transport but excluding taxis and hired cars.

Group 11: The supply of caravans and houseboats (but not to the supply of holiday accommodation within them).

Group 12: Gold.

Group 13: Bank notes.

Group 14: Drugs, medicines and surgical supplies.

Group 15: Imports and exports.

Group 16: Charities.

Group 17: Clothing and footwear for children.

Services exempt from VAT

Those supplies which are exempt from VAT are defined in Schedule 6 of the Value Added Tax Act 1983. The Treasury may alter the categories of supplies which are exempt at any time by making a Treasury Order (VATA 1983 s.17(2)).

The idea of exemption can be misleading. On the face of it exemption suggests that the activity and therefore the taxable person are to be free of the prescribed taxes. But with VAT this is not so. Since a person making exempt supplies cannot charge output tax on the goods and services that he sells she/he is equally prevented from recovering the input tax on the goods and services which she/he has purchased. She/he is therefore in the position of having to meet the charge to input tax her/himself.

The following are the present categories of exemption:

Group 1: The grant, assignment or surrender of any interest in or right over land or of any licence to occupy land but excluding the supply of:
(a) a freehold interest in a new or partly completed building provided that it is neither designed or intended to be used for residential or charitable purposes;
(b) a freehold interest in a new or partly completed civil engineering work;
(c) accommodation in a hotel, inn, boarding house or similar establishment;
(d) holiday accommodation in a house, flat, caravan or houseboat;
(e) facilities for camping, car parking, mooring boats or hangar facilities for aircraft;
(f) sporting rights;
(g) rights to sell or remove standing timber;
(h) space at an exhibition organized for the advertising of goods or services;
(i) facilities for sport, physical recreation or entertainment.
Group 2: Insurance
Group 3: Postal services (excluding telex and telephone services).
Group 4: Betting, gaming and lotteries.
Group 5: Finance, i.e. normal banking operations.
Group 6: Education by a school or UK university where this is done other than for profit and training for a profession or employment.
Group 7: Health services.
Group 8 Burial and cremation services.
Group 9: Services to members of trade unions and professions;
Group 10: Sports competitions.

Group 11: Works of art.

Group 12: Fund-raising events by charities and non-profit making bodies.

SUPPLIES OF INTERESTS IN LAND

Introduction

As we have seen, the supplies of interests in land are generally exempt (VATA 1983 Sch.6, Group 1). There are some exceptions to these exemptions and the more important ones are given particular consideration here, following an explanation of the option to tax certain supplies of land and buildings.

Option to tax

As from 1 August 1989 owners of interests in certain land and buildings have the right to elect to waive the exemption from VAT. The option to tax can be taken in respect of categories of supply falling within VATA 1983 Sch.6, Group 1 and includes sales of freeholds, grants of leases and licences to occupy premises. The consequence of the election will be for the landlord or owner to apply VAT at the standard rate and hence to enable the supplier of the interest in land to recover input tax. Such an election can be made in respect of any land or building but can only be in respect of the whole building and not of a part, (unless the owner's whole interest is limited to that part). Once the option to tax decision is made it is irrevocable.

Building and civil engineering works

The grant of a major interest in land will be zero-rated provided it is designed as one or more dwellings or is otherwise used for residential or charitable purposes. But such supplies apart the main category of supply of interests in land affected by VAT is the construction of new buildings which will be standard rated. Prior to 1 April 1989 supplies of such buildings, including a wide range of industrial and commercial buildings, were zero-rated (or standard rated if the construction was by way of refurbishment). Similarly all demolition and civil engineering goods and services carried out in the course of construction of these non-exempt buildings are also standard rated. A building is 'new' for the period of three years from the date of its completion. Buildings completed before 1 April 1989 are not new for the purpose of the three year rule. Developers with a legal commitment to supply a building or to receive the supply of building prior to 21 June 1988 and for which planning permission was obtained before 21 June 1988 will continue to be zero-rated provided that the supply takes place before 21 June 1993. These are transitional arrangements and any subsequent supplies in respect of such buildings will not be covered by these arrangements.

Civil engineering works such as the construction of roads and bridges are normally exempt. However, the supply is standard rated if the work is 'new'. Civil engineering work also includes infrastructure on building land such as access roads and sewers. Where bare freehold building land is sold (a supply which is exempt from VAT) then a standard rated supply may occur if the civil engineering work is very substantial. If the civil engineering work is minor then the supply will continue to be exempt. In some circumstances it may be necessary to apportion the supply between the exempt and standard rated elements.

Self-supply of building services

These services are deemed to be self-supplied by a trader in the course or furtherance of business and when the work is undertaken for other than monetary consideration. The services are deemed to be supplied by the trader for the purpose of her/his business, as, for example, where a developer constructs a building on her/his own land using her/his own employees. The value of the supply will then be the open market value of such services. The self-supply rules apply not only to new buildings but also to alterations, refurbishments and extensions. The charge to tax will only apply, however, when the value of the services so provided are in excess of £100 000 and if the supplies would have been charged at standard rate if supplied to a third party.

Hotel and holiday accommodation

While the provision of services in hotels and similar accommodation is standard rated, reductions are allowed for stays of greater than four weeks. As from the 29th consecutive day of letting the value of the accommodation supplied (i.e., the room rate as against say cleaning services) becomes exempt from VAT. This relief is subject to the amount remaining chargeable to tax being not less than 20% of the total value of services and accommodation supplied. The exception therefore carries a time bar which will vary with the proportion of services to accommodation charge.[17]

Camping and caravaning

The supply of facilities for camping and caravaning are only exempt when they are provided as permanent accommodation. This may be improbable with tented accommodation but quite likely with caravans in fixed locations supplied with utilities and other services. If the supply is in respect of mobile homes, holiday homes or other forms of non-permanent accommodation then there is no exemption.

Car parking

Car parking is exempt from VAT as a general rule. However there will often be occasions when car parking is offered to be supplied as part of an agreement to sell or lease property. In these circumstances the car parking would normally be considered to be part and parcel of the more significant supply and would then enjoy the status of the larger supply of an interest in land thus avoiding any possible need for apportionment.

The granting of sporting rights

The granting of any right to take game or fish is chargeable whether by lease, licence or other means. This will be true even where there is a larger right granted over land such as a tenancy of the whole land and although there may be no mention of them, the enjoyment of sporting rights are clearly seen to be a benefit arising from the lease. Where this occurs tax will be chargeable on that proportion of the consideration which is applicable to the sporting rights. In these circumstances separate valuations might be made.

Additional difficulties will occur where a landlord grants shooting rights to a syndicate. The ultimate tax position will depend upon the various administrative and financial arrangements between the parties. Sufficient to say that it has been held that where a landlord invited others to shoot with him on the basis of shared costs that the landlord had made taxable supplies, but since this was done in pursuit of pleasure and social enjoyment, and not in the way of trade, it was not therefore a business transaction for the purposes of VAT.[18]

Sale of building land

The sale of either the leasehold or freehold interests in bare land is exempt from a charge to VAT unless the vendor wishes to exercise the option to tax. If the vendor chooses not to opt to tax VAT may still be levied on the (historic) cost of the land when future development is completed.

Change of use and refurbishment

A trader is deemed to supply her/his interest in a building if it ceases to be used for a residential (or charitable) purpose after 1 April 1989 and provided that the change of use occurs within ten years of the completion of the building. The date of supply is the date when the change of use occurred. This could be the actual change of use or the date when planning permission for such a change of use was granted. The value of the supply will then be the aggregate value of the supplies in respect of the building made to the trader after 1 April 1989 and the supply will be chargeable at the standard rate.

So far as refurbishment of commercial property is concerned then it continues to be standard rated as it was before 1 April 1989. However, the refurbishment of listed buildings which had previously been zero-rated, is standard rated from 1 April 1989. There can be difficulties in distinguishing between 'alterations', 'refurbishment' and 'reconstruction'. Some aspects of the interpretation of the meaning of the relevant words is discussed below.

Sale of freehold

Formerly the freehold sale of a building was exempt and the freehold sale of old buildings will continue to be exempt (unless the vendor opts to tax). For these purposes an old building is one which was completed before 1 April 1989 and which was also fully occupied at that date. On the other hand the sale of a new freehold building is liable to VAT at the standard rate. For this purpose the definition of a new building is one which was completed after 1 April 1989 and which was completed less than three years before the date of sale. Here completion will refer to the date of issue of a certificate of practical completion or alternatively the date when the building was first fully occupied. It should be noted first that the date of practical completion may precede certain finishing works, and secondly that the date of occupation may refer to the date of an agreement to occupy rather than to the date of physical occupation.

Sale of leasehold

When a building is leased, whether on a short or a long lease, the sale of the interest is exempt, unless the vendor opts to tax, this is so whether the building is an old or a new one. The option to tax can occur at any time (i.e. not just at the commencement of a lease) and is in respect of the whole building. Once the option to tax has been exercised it is irrevocable and any subsequent sale of a leasehold interest in the building will attract VAT at the standard rate.

Occupational leases

Since 1 August 1989 landlords have had the choice of continuing to receive rents as an exempt supply or to tax rents received from tenants with VAT at the standard rate. Since the rents so received are a taxable supply the landlord is able to recover input tax on any taxable supplies made in respect of the building on which the option has been taken. Those tenants who are standard rated will not suffer any increase in their occupation costs since they themselves will be able to recover input tax. Those tenants who are fully or partially exempt will suffer an increase in their occupation costs since they will be unable to recover input tax. The decision of the landlord to exercise the option to tax will depend very much therefore on the circumstances of each case.

SERVICES SUPPLIED IN CONNECTION WITH LAND

Where property is offered for lease it is usually the case that additional services will be offered. These services may be supplied on a common basis between a number of tenants. They may be a specific condition of the lease or an optional service. In addition the granting of the leases and management of the property may bring about the need for services to the person granting the lease and responsible for ongoing obligations under the lease. Depending upon the circumstances the tax position will vary.

Services provided by a landlord

Particularly where commercial premises are offered for lease (but also in the case of flatted residential property), common services will be supplied. The normal practice will be for the leases to provide that the landlord will supply certain specified common services, such as the maintenance and cleansing of the common parts. Where these charges are not included in the rent they are usually collected as service charges. Although additional to rent they are nevertheless part of the consideration received by the landlord in return for a right to occupy the premises. A more detailed examination of the taxation position of services provided by a landlord follows:

a) Insurance
The insurance of the building will fall within the general category of insurance specified elsewhere in Schedule 6 of the VAT Act of 1983 and any costs therefore will be exempt.

b) Common services
Whether a tenant pays a service charge or whether he pays additional rent for common service supplied and the maintenance of the building as a whole then such payment is considered to be rent in fact. Hence the

amounts involved are exempt in a non-elected building and standard rated in an elected building where the buildings are commercial. Where these services are exempt, and there is a general exemption for the granting of interests in land, then this inevitably means that the landlord has to bear the input taxes on the costs he has incurred. It is not possible for him to pass the tax on to the tenants. This may be unfortunate where the tenants of commercial property may be taxable persons who could therefore absorb the tax by claiming credits. Landlords are likely to opt to tax in these circumstances.

c) Specific charges

Services may be supplied by a tenant in respect of services such as cleaning, heating or lighting and in respect of the premises which the tenant occupies exclusively and not in common with others, and not charged separately. These services are then considered to be separate from any grant of an interest in land and will be charged as taxable supplies. The rate of tax will then depend upon the nature of the service or goods supplied so that cleaning, for example, would be taxed at the standard rate, while heating or lighting would be zero-rated.

Services provided for a landlord

A landlord who arranges for the disposal of interests in land will himself require services in connection with the arrangement of the contracts, and depending on the circumstances he may have to incur ongoing management expenses.

a) Landlord's professional charges

Where services have to be provided in respect of financial matters, (say the arrangement of a mortgage) then there is the general exemption afforded in Group 5 of Schedule 6. Services such as those provided by a Chartered Surveyor or solicitor are chargeable to tax at the standard rate. It should be noted, however, that normal practice in respect of the granting of commercial leases is for the tenant to meet the landlord's costs in connection with the preparation of the lease. Since this cost is not a supply to the tenant s/he need not add VAT to the payment s/he makes to the landlord for this service.

b) Management services

The cost of management services which will include the collection of rents and administrative work in connection with the ongoing maintenance and management of the property will be chargeable at the standard rate.

SUPPLIES OF CONSTRUCTION

Introduction

Before April 1989 construction work was zero-rated and refurbishment work was standard rated. The distinction between construction and refurbishment was therefore of some importance and this need to distinguish between these two activities is now less important given that all new construction and refurbishment is standard rated. It is still necessary to know how those terms are defined for the purpose of VAT and a resumé of some aspects of the supply of services for construction therefore follows.

The phrase 'the person constructing the building' can be important. It is not apparently necessary for a person to undertake the actual construction himself. For example a developer may instruct a builder to construct a building on land which the developer owns. Provided the developer retains control over the development he will be the constructor of the building. The test of such control may be evidenced by a contract between the parties whereby the contractor undertakes to provide the completed development to the satisfaction of the developer who otherwise retains control over the planning and design of the development.

As to what is a 'building' the legislation is silent. The definition of a building is important for all taxable situations. For the purpose of VAT it would appear that a building must be a structure with foundations. It may be fixed to these foundations but a temporary structure erected on permanent foundations is a building.

The word 'construction' includes the entire process of construction including any activity from clearing of the site to the formal completion of the building.

'Reconstruction' on the other hand has proved more difficult to define and the distinction between 'construction' and 'reconstruction' is not always an easy one. Before 1984 a simple view was adopted. The Customs and Excise office accepted that the test could be one of cost. Essentially major reconstruction could be regarded as construction if the cost of such reconstruction were more than half the cost that it would have been to construct such a building *de novo*. The position now is that the Customs and Excise seek an ordinary meaning to the words 'reconstructing, altering or enlarging an existing building' rather than an arbitrary test. Hence the modern test is to establish whether a new building is being constructed or whether works are being carried out to a building which is already in existence.[19]

In clarification of this position the Customs and Excise view the following situations as ones where 'construction' is being undertaken:

1. constructing a new building on the foundations of an old building;
2. constructing a new building where only a single wall of the old building remains;
3. building a separate house on the gable wall of an existing house;
4. building a new house by infilling a site between existing terraced houses where the original house had been demolished;
5. new construction within a large building where there is no enlargement of the existing building.

The Customs and Excise view the following situations as ones where 'reconstruction' is being undertaken:

1. building operations making use of the shell of an existing building;
2. building operations where, in addition to a single wall remaining there may be internal features included in the construction;
3. building an additional flat on top of a block of existing flats.

It should be noted that there are special provisions relating to the reconstruction of buildings and monuments of architectural or historic importance. Where such a listed building has been 'substantially reconstructed' (i.e. more than 60% of the cost of the work is attributable to alterations) then such work may be considered to be 'construction'.[20] (VATA 1983 Sch.5, Group 8A and C&E Leaflet No.708/1/4).

Customs and Excise appear to consider that building works are only 'alterations' if, as a matter of fact, they actually alter the building in some important respect. If a building is altered in such a way that the effect is to change the complete building into another type of building then it is an 'alteration'. Alteration may be categorized by intensification of existing use, say by the conversion of a single dwelling into a number of flats.

As with 'construction' and 'reconstruction' there can be a blurring if distinction between 'alteration' and 'repair'. In every case the question of whether work is alteration or repair will be one of fact. However base tenets of what constitutes repair can be kept in mind. Repair and maintenance clearly envisages that the property is to be retained in (or, if necessary, restored to) the condition which obtained at the time it was constructed. If necessary this concept can be extended to suggest that the objective of repair and maintenance is to carry out such work as is necessary in order to preserve the value of the property. These ideas admit the notion that old parts or fittings may be replaced with new where it would otherwise be absurd to replace with old. Hence the VAT Tribunal has found that the replacement of an old kitchen and its original fittings with a modern fitted kitchen was repair and not alteration.[21]

'Civil engineering works' are broadly those works of a permanent nature which form part of the 'land' (i.e. land and buildings) but which are not repair and maintenance works.[22] The kind of works which have been recognized as civil engineering works include the construction of a septic tank,[23] the construction of a swimming pool,[24] and the excavation of land and the building of a retaining wall.[25] However where the work is within the grounds or garden of a private residence it is treated as residential construction. Agricultural land drainage and water supplies may be engineering works. In order to be recognized as civil engineering works the land drainage system must comprise of (mainly) permanent under drainage and it must be self contained (i.e. it must provide drainage for a distinctive physical area – a field or a farm).[26] Similarly a water supply system for irrigation or for the use of livestock will qualify as civil engineering works if it consists mainly of underground pipes.

VALUE ADDED TAX AND LAND VALUE

So far as the valuation of land is concerned VAT raises more questions as to what is liable to taxation where land is concerned than it does about what its value may be. Generally the legislation implementing VAT is indifferent to the price at which commodities change hands.

As we have seen only two matters regarding the valuation of goods or services have concerned the legislators. First, where a taxable supply is for a consideration in money then that sum shall be held to be tax inclusive. (VATA 1983 s.10(2)). Secondly, where the supply of goods is not for a consideration in money or where it is partly for consideration in money then the monetary value is to be derived by reference to the 'open market value' (VATA 1983 s.10(3) and (5)). The concept of 'open market value' is long standing in tax legislation and case law and an attempt to understand its definition is made in Chapter 9.

In the meantime it may be noted that VAT is also relatively indifferent to the treatment of capital and income. This is in marked contrast to other areas of taxation which affect land. For most taxation matters affecting land a careful distinction has to be made between revenue and capital elements in any given situation.

However, VAT is only concerned with the supply of goods and services. The supply of a major interest in land is a good. Any interest in land which is not a major interest is not a good and therefore its supply is a service. But more importantly, whether a good or a service, the supply will be taxed on the consideration whether that is a lump sum, a series of payments (i.e. a continuous supply of goods and services) or a combination of both. So far as VAT is concerned the identification of income streams, capital gains or windfall profits is not a matter of any importance. The main consideration will be as to whether a supply of land should be standard-rated, zero-rated or exempt. This would only be of interest so far as land was concerned if it were decided to bring in different rates of tax on goods and services. Such an event would have a major impact given the rather arbitrary definition of 'major interest' in land and the consequential division of interests in land between goods and services.

However there are indirect effects from the incidence of VAT on supplies of land which may give rise to problems in respect of the value of land. Following the introduction of the standard rate of tax on new construction in 1989 there is at least the possibility that a two tier rental market may be created in certain office centres. Since those companies which are tax exempt are primarily from the financial sector, public administration, education, health and recreation then market distortions may occur in locations where the office market is significantly dependant upon such business sector occupiers. Where this occurs exempt tenants will tend to bid for space in non-elected buildings because their occupation costs will be increased in elected buildings where they will be required to pay rent plus VAT, which they will not be able to recover. This preference for non-elected buildings by tax exempt tenants will theoretically hold until rents are bid to 15% over market rents at which point tax exempt consumers will become indifferent to taking space in elected or non-elected buildings.

From the point of view of the standard-rated occupiers of office space then they will be discouraged from seeking space in non-elected buildings because of the aggressive bidding of the tax exempt tenants. Rather the demand for space by standard-rated tenants will tend to converge on that part of the stock of available office space which is housed in elected buildings and where they will be able to recover input tax.

The logical outcome of this process would be a two-tier market in office rents. The two different levels of rents would be physically segregated between non-elected buildings occupied by tax exempt tenants and elected buildings occupied by standard-rated tenants. The former buildings would have a level of rent up to 15% above market rents while the latter would be let at market rents on which VAT will be charged at the standard rate.

While this scenario of a two-tier rental market makes financial sense for the consumers of office space the actual office market will be modified by other factors. A major factor will be the rate at which the VAT element will filter through the stock of properties in the market. To begin with the VAT element will enter the property market through developers and landlords who will incur VAT on new developments and refurbished properties. Clearly there will be a strong incentive for the owners of such properties to recover VAT charges on their properties by opting to tax the rents of tenants on occupation leases or on the sale of these properties. But the incidence of new development will mean only a marginal and gradual increase in the stock of elected offices. The provision of the option to tax will mean that existing properties may also be brought into the taxation net. However, the financial incentive on the owners of these properties will be not to exercise their option to charge VAT on rents. If they do so they are no better off, but if the building is a non-elected one then they can hope to attract tax exempt tenants who are willing to pay a premium rent. When this occurs the owner's return on his investment in office property will increase through no extra effort on her/his part or the acceptance of any higher level of risk.

It would appear therefore that there is a coincidence of financial motive between both those who demand and those who supply office space which will tend to support a two-tier market and to resist the longer term trend towards the universal application of VAT throughout all commercial property. Since most business are standard rated and are therefore able to recover VAT, the taxation of both landlords and tenants will be neutral with the tax actually borne by the ultimate consumers of goods and services produced from within these properties. Where exempt occupiers are a minority in any given location there is unlikely to be the necessary competitive bidding to establish premium rents in non-elected buildings, but where exempt occupiers are significant as in the City of London or the New Town of Edinburgh a two-tier market could develop even allowing for the market imperfections which will inevitably act to reduce the differences in rental levels between the two tiers. In any event, the long run will re-establish a unitary market, but where these divergent interests exist in the short-term then the value of property investments may be indirectly affected by VAT as a charge on new commercial developments and on rents.

STAMP DUTY

Introduction

Stamp Duties are among the longer-established forms of taxation having first been imposed as a tax on documents in 1694. The principal legislation governing duties on documents is the Stamp Act 1891, (together with the Stamp Duties Management Act 1891) and various amending Finance and Revenue Acts. The most significant of these has been the Finance Act 1985 which introduced considerable changes to simplify duties and bring them up to date.

Although stamp duties have altered very little over the years there has been a recent movement towards examining any inhibiting effects there may be in reducing the efficiency of the economy by maintaining fiscal barriers to the transfer of assets. Government has taken specific measures in the recent past to reduce the incidence of taxation via Stamp Duty. In the Finance Act 1984 the rate of duty on the sales of land, buildings and other property was halved (from a long standing rate of 2%) to 1% on transfer or sales over £30 000. This was accompanied by a raising of the threshold value on which stamp duty became payable on such transactions. The same Act reduced duty on sales of shares and other property by the same amount and the Finance Act 1986 reduced the duty on sales of shares again to 0.5% with effect from 27 October in that year. (Therefore co-terminal with the 'big-bang' in the provision of financial services which occurred on that date). In the budget of 1990 it was announced that Stamp Duty on the transfer of securities would be abolished towards the end of the financial year 1991/92.

Nevertheless Stamp Duty remains a small but important and growing source of revenue for the Government (£2127m in 1989 or 1.67% of all Government revenues) and it is very cheap to collect. It remains therefore as a tax on land transactions although since it is a tax upon instruments (written documents) it is avoidable where verbal contracts are deemed to be sufficient. This having been said, the law unequivocally requires certain land transactions to be made in writing thus inescapably making them subject to Stamp Duty.

The nature of stamp duties

Stamp Duties are levied under what are known as Heads of Charge (SA 1891 s.1). They are governed only by statute and this means that no document can be charged with stamp duty unless it is specifically embraced by an Act of Parliament. The Heads of Charge are therefore clearly important. The duties specified (SA 1891 Sch.1) are of two kinds, i.e., 'fixed' and 'ad valorem'. As implied, fixed duties do not alter. Ad valorem duties on the other hand alter both with the amount of the consideration and with the structure of the scales laid down for them. Since the Finance Act 1985 those fixed duties which remain are only there to enable to Stamp Office to scrutinize documents that may be liable to Stamp Duty.

With only minor exceptions the duties for land transactions are ad valorem.[27]

The main heads of charge which affect land and on which ad valorem rates of tax are payable include the following.

1. The conveyance or transfer on sale of property.

2. The lease of land.
3. The conveyance or transfer operating by way of voluntary disposition (F(1909–10)A 1910, s.74(1)).

Administration

Stamp Duty is administered by the Commissioners of Inland Revenue. As we have seen the instruments to be taxed are specified by Act of Parliament (SA 1891 s.1). However, determining the amount to be paid in tax may be less certain. The administrative procedures to allow for such determination require that a document be submitted to the Inland Revenue for 'adjudication'. There is a right of appeal against an Inland Revenue decision on adjudication to the Chancery Division (or High Court in Scotland) by way of stated case, from where, if necessary, the matter can be taken to the House of Lords.

The adjudication process therefore enables the correct amount of duty to be established by the Inland Revenue. In determining the value on a transfer or sale where *ad valorem* duty is chargeable but in conditions of uncertainty certain contingent rules apply.

1. If it is not possible to fix a minimum or maximum amount on the subject of transfer or sale then no *ad valorem* duty can be charged.[27] On the other hand if a figure is fixed it may be taken, even although it may be varied up or down in an unforeseen way.[28]
2. If a minimum figure is fixed then *ad valorem* duty can be calculated on the figure. A fixed duty (50p) can then be charged on the possibility of the larger amount being realized[29] (SA 1891 s.4(b)).
3. If a maximum figure can be determined then *ad valorem* duty will be calculated with reference to that figure even although the consideration which is eventually paid may be less.[30,31,32]

Exemptions

There are some general and particular exemptions from Stamp Duty afforded by the legislation including one or two minor ones which affect land.[33] The most significant current exemption is for conveyances or transfer on sale. Voluntary dispositions and leases are exempt if made to the Historic Buildings and Monuments Commission which, like other persons or trusts established for charitable purposes enjoys this relief (FA 1983 s.45(3)). However, it will be necessary for a claim for relief to be adjudicated (FA 1982 s.129).

The legislation defines 'conveyance' (SA 1891 s.54) and this requires a transfer of 'property' consequent on a 'sale'. The Act does not define 'property', 'sale' or 'consideration' but case law has determined their meaning and purpose over the years. As might be imagined a degree of consent between the parties would be expected to be a necessary condition. However, it has been held that an acquisition using powers of compulsory acquisition is a sale.[34]

Where the consideration is by way of cash its determination is not difficult but difficulties may arise when a transfer is made and where there may be doubt as to whether the consideration is adequate. It is important that the instrument is stamped on the value of consideration and that the consideration is adequate. If the consideration is not adequate the Inland Revenue can later object that the instrument has not been properly stamped since it is a voluntary disposition at under-value. While this is a sufficient difficulty in itself it also carries the risk that the purchaser can claim that the title is defective.

The rates of duty for conveyances or transfer on sale are 1% and for most categories no duty is payable if the value is no more than £30 000.

Conveyance of domestic property

At the time of the introduction of the Stamp Act 1891 the purpose of levying taxes on transfers of capital was clearly a measure designed to tax wealth which was concentrated in a relatively small proportion of the population, particularly so in the case of land. Fewer than 10% of the population owned their own houses and since Stamp Duties on leases were payable by the landlord it was clear that the vast majority of the population would be quite unaffected by their introduction.

The trend towards home ownership, particularly since the Second World War, together with general inflation and inflation in house prices in particular has progressively brought about a position whereby the majority of the population own or live in owner-occupied property and the majority of domestic properties transferred have a market value in excess of the Stamp Duty threshold for *ad valorem* rates. Bodies such as the Royal Institution of Chartered Surveyors have regularly petitioned the Chancellor of the day on this aspect of taxation on domestic properties. Critics have drawn attention to structural inadequacies in the schedule with consequences for the impact of taxation which are held to result in maldistribution. An obvious criticism is that the tax is not progressive after relatively low capital values so that the rich, *pro rata*, pay much less in tax when they sell very expensive residences.

Example 5.2

A & B sell houses for amounts of £30 000 and £36 000. Their liability to Stamp Duty is as follows.

	Consideration £	Rate of tax	Stamp Duty £
A	30 000	nil	nil
B	36 000	1%	420

Note: A pays no tax at all on £30 000 but B pays tax not only on the amount by which the purchase price exceeds £30 000 but on the whole amount.

Land transactions are not often subject to vague and uncertain contingency amounts being paid, but it can happen and when it does the rules expressed above will apply.

The adjudication process permits any person to require the Commissioners to indicate what the liability to duty (and the amount) may be on any given transaction. When such an opinion is sought and given the investment *may* be stamped with the amount of duty which has been determined. In addition an adjudication stamp may be impressed on the instrument affirming the amount of duty chargeable, or possibly indicating that no duty is chargeable. After having received an opinion from the Commissioners on a matter of adjudication, however, there is no *obligation* to pay the duty and the instrument may not be stamped.

Nevertheless, adjudication is compulsory in certain cases, including the following.

1. Voluntary disposition (SA 1891 s.12(2)).

2. Conveyance in contemplation of sale (FA 1965 s.90).

It is also required, *inter alia*, when exemption is claimed in respect of the following.

1. Maintenance funds for historic buildings (FA 1980 s.98).
2. Conveyances, transfer or leases to charities or the National Heritage Memorial Fund of the Historic Buildings and Monuments Commission (FA 1982 s.129 FA 1983 s.46).

In any event the adjudication process will be the first stage in challenging the Stamp Duty Office as to the appropriate amount of consideration for the purpose of the tax.

Prior to 1984 the threshold was lower, but average house prices have rapidly outstripped the threshold level of £30 000 which has obtained since then, hence fuelling further demands for the reform or removal of the tax.

However some of the arguments against its removal include the observation that since home ownership is now reaching saturation point, the tax has not proved to be, nor will it be an inhibiter to the achievement of widespread home ownership. Secondly, it is argued that removal of the tax is only likely to result in increases in the prices of those properties which suffer its incidence.

Leases

Stamp Duty is payable under the head of charge 'lease on tax'.[35] The tax is leviable on the granting of a lease for a term of one year or more. 'Lease' is not defined in the Act but the word has been interpreted as excluding a licence.[36] The lease must be of heritable properties so that leases of non-heritable and moveable property are not included. The charge is *ad valorem* on the rent (and on any premium).

Leases which are not for a definite term of one year or more are covered by other headings of charge in the first Schedule and in these cases the amount of duty is fixed. These include furnished tenancies for less than a year, (fixed duty of £1) and other leases not otherwise specified (fixed duty of £2).

Rent

The legislation does not define 'rent' but the courts have defined it as the compensation paid to the landlord for the sacrifice of exclusive possession of the land.[37] From the point of view of the land valuer the Stamp Duty Act 1891 uses some unfortunate terminology so that reference is made to the *average* rate of the yearly rent payable under the lease. Hence if a lease is for £10 000 for seven years, £20 000 for the next seven years and £30 000 for the final seven years the lease will be treated as a twenty-one year lease at an annual rental of £20 000 and taxed accordingly. Where the rent is variable, (for example where there is a reference to review of the rent to market value or not less than rent passing at the end of a specified period), then the contingency rules will apply. In this case the minimum amount only is certain and it is that on which Stamp Duty will be paid.

Where the landlord provides common or particular services they will be liable to Stamp Duty unless they are specifically reserved as rent.

The importance of establishing the amount of the rent which is dutiable lies in the fact that duty is charged, *ad valorem*, on such rent. The rate of duty will in turn be affected by the length of the lease.

Duration of the lease

It is important to determine the length of the lease when considering the incidence of stamp duty since the rate rises steeply with the duration of the lease. As we have seen the length of the lease is also of importance in deciding the heading of charge.

Statute requires therefore that the term of the lease be decided, and if a lease is for a fixed term, but with qualifying clauses or conditions, then it will be necessary to determine the length of the lease which satisfies the requirements of the legislation. For example a lease which is for a fixed term with the right to hold for an unspecified period at the end of the term, then the lease will be classified as one for a fixed term the length being equal to the length of the fixed term plus the minimum period of time thereafter following which the lease can be terminated (FA 1963 s.56(3)). On the other hand a lease for a fixed term terminable on some future event will be treated as a lease for a definite period equating with the fixed term.[38] In the same way leases for a fixed term but with options to review for further fixed terms are treated as leases for the original fixed term only.[39]

Duty payable on rents
The duty is charged on the rent passing on an *ad valorem* basis, the rate increasing as the length of the lease increases.

Table 5.1. Stamp Duty payable on leases

Amount of Rents	Term not exceeding 7 years or indefinite	Term not exceeding 35 years but exceeding 7 years	Term Not exceeding 100 years but exceeding 35 years	Term exceeding 100 years
		£	£	£
£5 or less	nil	0.10	0.60	1.20
Over £5 and up to £10	nil	0.20	1.20	2.40
Over £10 and up to £15	nil	0.30	1.80	3.60
Over £15 and up to £20	nil	0.40	2.40	4.80
Over £20 and up to £25	nil	0.50	3.00	6.00
Over £25 and up to £50	nil	1.00	6.00	12.00
Over £50 and up to £75	nil	1.50	9.00	18.00
Over £75 and up to £100	nil	2.00	12.00	24.00
Over £100 and up to £150	nil	3.00	18.00	36.00
Over £150 and up to £200	nil	4.00	24.00	48.00
Over £200 and up to £250	nil	5.00	30.00	60.00
Over £250 and up to £300	nil	6.00	36.00	72.00
Over £300 and up to £350	nil	7.00	42.00	84.00
Over £350 and up to £400	nil	8.00	48.00	96.00
Over £400 and up to £450	nil	9.00	54.00	108.00
Over £450 and up to £500	nil	10.00	60.00	120.00
Over £500: for every £50 (or fraction of £50)	(1)	1.00	6.00	12.00

Note: Leases not exceeding 7 years, or for an indefinite term are not chargeable to duty where this does not exceed £500 pa. Should the amount of rent be in excess of £500 pa the duty is £1 for a lease of furnished rental accommodation for a fixed term of less than 1 year, or in all other cases 50p per £50 rent (or fraction of £50)

Duty payable on premiums

Where a premium is paid in respect of the granting of a lease then duty is chargeable *ad valorem* at the same rates obtaining on a conveyance or transfer on sale. This duty will be payable in addition to the duty payable on rents.

Voluntary disposition

Until March 1985 a conveyance or transfer of land carried out by way of 'voluntary disposition' fell to be charged with stamp duty just as if it were a normal conveyance or transfer on sale (F(1909–10) A 1910, s.74(1)). The duty payable was also the same as that which applies to a conveyance or transfer on sale (i.e., *ad valorem*), and an adjudication stamp was required.

The matter of the consideration and the chargeable amount required special treatment, however. For example, A could sell to B land and buildings worth X in the open market for the lesser sum of Y. There are two possible taxable outcomes. The instrument conveying the land could be taxed as a sale taking account of the consideration paid. Alternatively the transfer could be taxed as a gift, the consideration being the difference net between the full value of the land and the price paid out on the full value of the property.[40] The election of treatment rests with the Inland Revenue and one supposes that they choose that taxable event which produces the greatest revenue to the Nation!

Since March 1985, however, there is no longer any *ad valorem* duty on gifts and the need for adjudication has been removed. Gifts now suffer a fixed duty of 50p only.

Finally, it should be noted that Stamp Duty is payable on the total consideration for a transfer of an interest in land including the charge to VAT. Most properties will be conveyed on taxable supplies and hence the amount paid in VAT will be recoverable. However, where properties are conveyed as exempt supplies no VAT will be recoverable. The payment of Stamp Duty on the non-recoverable VAT gives rise to double taxation.

NOTES: CHAPTER FIVE – TAXES ON EXPENDITURE

1. Turnover taxes on the gross earnings of businesses have been widespread in Europe where evasion of taxes on income and profits had been a problem.
2. The wording of the Act does not always match the wording of the directive. Where this occurs reliance should be placed on the directive which takes precedence in the United Kingdom.
3. While VAT is leviable in the Isle of Man it has its own administration and collection machinery.

4. VATA 1983, Sch.1 para.1 and S.I's in force.
5. The amount is variable and the limit was last set as from 21 March 1990.
6. VATA 1983, Sch.1 para.2 and S.I's in force.
7. *Customs and Excise Commissioners* v. *Morrison's Academy Boarding Houses Association* [1978] S.T.C. 1; see also VAT 1983 Secs.47(1) and 48(2).
8. VAT is different from other taxes in this and other ways. It is a standard tax which is not broken down into its several sources and it is also self-assessed.
9. A 'major interest' in land is defined as a freehold (or heritable) interest or a lease for more than twenty-one years (VATA 1983 Sec.48(1) and Sch.2 para.4).
10. The difference between the transfer of assets here and those deemed to be included in 3(c) is that because the business is transferred as a going concern the purchaser becomes taxable (if he is not already) upon acquiring the business.
11. Note that this provision cannot apply to the supply of services, where if there is no consideration there can be no supply (VATA 1983, Sec.3(2)(b)).
12. VATA 1983 Sch.4.
13. VATA 1983 Sch.3.
14. This rate has obtained since June 1979, however the Treasury has power to change the rate by 25% of its obtaining level by use of a statutory instrument (VATA 1983 Sec.9(1) and (2)).
15. VATA 1983, Sch.6.
16. In *Geoffrey E. Snushall* v. *Customs and Excise Commissioners* [1982] S.T.C. 537 it was held that a property guide produced by a firm of surveyors was not a newspaper.
17. Note, however, that the supply of meals and drinks is always taxable at the standard rate (VATA 1983 Sch.4 para.9).
18. *Lord Fisher* v. *Customs and Excise Commissioners* [1979] V.A.T.T.R. 227.
19. Following the finding in *Great Shelford Free Church* v. *Commissioners of Customs and Excise* [1985] V.A.T.T.R.; [1987] S.T.C. 249 where an extension to a church was held to be a new building and therefore zero and not standard rated the law was changed by the approval of Parliament of SI 1987 No. 1072 which clarified the criteria for distinguishing between enlargement of an existing building and new construction.
20. As an alternative to this condition the reconstructed building must incorporate no more of the original building than the external walls and other external features recognized as of national interest.
21. *Multyflex Kitchens Ltd* v. *C and E Commissioners*. C.A.R./75/213.
22. Also excluded is exploratory work.
23. *Murray* v. *C & E Commissioners* L.O.N./74/122.
24. *En-tout-cas Ltd* v. *C & E Commissioners* [1973] 1 Q.A.T. T.R. 101.
25. *Skeffington* v. *C & E Commissioners*, L.O.N./74/35.
26. Permanent under drainage normally means the laying of pipes at least two feet below the surface.
27. The exceptions are 'conveyance or transfer of any other kind not hereinbefore described'. (An example might be a conveyance in considerations of marriage – which is not a sale, and leases of small furnished lettings.) For voluntary dispositions completed after 25 March 1985 the duty changed from 'ad valorem' to a fixed rate.
28. *Underground Electric Railways* v. *IRC* [1916] 1 K.B. 306 CA.
29. *Independent Television Authority and Associated Rediffusion Ltd* v. *IRC* [1961] A.C. 427.
30. *Jones* v. *IRC* (1895) 1 Q.B. 484.

31. *Underground Electric Railways* v. *IRC* (1906) A.C. 21.
32. The logic of this rule means that where both a minimum and a maximum figure can be set then *ad valorem* duty will be calculated with reference to the larger amount.
33. For example, the amount of duty payable on leases of allotments is limited to 50p unless the rent exceeds 50p or a premium is paid (Land Settlement (Facilities) Act 1919 s.21).
34. *IRC* v. *Glasgow and South Western Railway* [1887] 12 A.C. 315, see also *Kirkness* v. *John Hudson and Co.* [1955] A.C. 696.
35. SA 1981, Sch.1(3).
36. *Addiscombe Garden Estates Ltd* v. *Crabbe* [1958] 1 Q.B. 513 CA.
37. *Hill* v. *Booth* [1930] 1 K.B. 381.
38. *Earl Mount Edgecumbe* v. *IRC* [1911] 2 K.B. 24.
39. *Hand* v. *Hall* [1877] 2 Ex. D. 355.
40. *Baker* v. *IRC* [1924] A.C. 270.

Chapter Six

INHERITANCE TAX

INTRODUCTION

Inheritance Tax is a tax made on what are known as 'chargeable transfers' made by a taxable person during the person's life and/or on the person's death.

Inheritance Tax was previously known as capital transfer tax and before that as Estate Duty ('death duty'). The tax is levied on capital and is intended to be progressive and distributive.

Estate duty ceased to apply to deaths occurring after 12 March 1975, after which capital transfer tax applied.[1] Capital Transfer Tax was more vigorous in its pursuit of taxable capital than Estate Duty, particularly in respect of the taxation of transfer of capital made during the lifetime of the taxable person. It also made the taxation of capital at death less avoidable than had been the case under the former Estate Duty provisions.

Following the change of government in 1979 substantial changes were made to Capital Transfer Tax in subsequent Finance Acts. These changes were consolidated in the Capital Transfer Tax Act 1984, now known as the Inheritance Tax Act (ITA). The Finance Act 1986 introduced 'potentially exempt transfers' which meant that since then gifts made will not be liable to inheritance tax unless the donor dies within seven years of the gift being made.

This latest change has re-introduced the possibility of the tax being avoidable for the planning-conscious taxpayer and brings the potential incidence of the tax broadly back to where it was under estate duty, although there can still be liability for lifetime transfers.

Property chargeable

Inheritance Tax is charged on the value transferred by a 'chargeable transfer' (ITA 1984 s.1). Subject to various exemptions and reliefs, therefore, the tax will be charged on the decrease in the tax payer's assets as a consequence of the transfer of ownership.

The Act defines a chargeable transfer as any 'transfer of value' made by an individual other than transfers which are exempt (ITA 1984 s.2(1)).

A transfer of value is any disposition made by a tax payer as a consequence of which the value of the remaining property is diminished. The value transferred therefore is the amount by which the value of the individual's estate has been reduced. For this purpose an individual is treated as having made a notional transfer of her/his estate at a time immediately before her/his death.

In addressing the question of what may be included in the person's estate the Act envisages that this will comprise the aggregate of all of the property to which that person is beneficially entitled (ITA 1984 ss.5(1) and 49(1)).

Inheritance tax applies to all of the taxpayer's property if that person is domiciled in the United Kingdom, wherever that property may be located.

POTENTIALLY EXEMPT TRANSFERS (ITA 1984 s.3A(4), FA 1986 Sch.19 para.1 and F(No.2)A 1987 s.96 and Sch.7)

After 17 March 1986 any gift or disposition which would normally be classified as a chargeable transfer becomes a 'potentially exempt transfer'. This means that inheritance tax will not be paid on these gifts unless the transferor dies within seven years of the disposition. Hence, a potentially exempt transfer made seven years (or more) before the transferor dies becomes exempt, but a potentially exempt transfer made within that seven year period becomes a chargeable transfer. It is assumed that a potentially exempt transfer is an exempt transfer until either the seven years runs out or the transferor dies.

AMOUNT OF TAX (ITA 1984 s.7 and FA 1988 s.136)

The nature of inheritance tax requires that chargeable transfers are accumulated so that the tax is charged on a cumulative total of chargeable lifetime transfers plus the estate which is notionally disposed of immediately before death. Under the former capital transfer tax the cumulative total would indeed represent an aggregation of chargeable transfers over a ten year period prior to death. However, following the introduction of potentially exempt transfers in the Finance Act 1986 a seven year limitation on accumulation applies from 18 March 1986.

Inheritance tax is charged at a single rate (40%) on transfers on death since the Finance Act 1988, with half that rate (20%) applying to lifetime transfers. Previously it had been charged at progressive rates. The present simplified basis of charge is shown in Table 6.1.

GROSSING-UP

If a taxpayer makes a chargeable transfer then the amount of tax will be levied on the loss to the donor rather than on the benefit to the donee. But the donor may choose to pay the tax himself (based on the value transferred) or have the donee pay the tax which would then have to be

Table 6.1. Transfers on death and during lifetime after 6 April 1990

Transfers on death after 6 April 1990

| | *Tax on transfer* | | |
Slice of cumulative chargeable transfers	*Cumulative total*	*Rate*	*Tax payable*
£	£	percent	£
< 128 000	0–128 000	nil	nil
> 128 000		40	40% of amount > 128 000

| | *Grossing-up of transfer (where applicable)* |
Net transfer	*Tax payable*
£	£
< 128 000	nil
> 128 000	nil + 2/3 of amount > 118 000

Transfers during lifetime after 6 April 1990
(i.e. lifetime transfers which are not potentially exempt transfers)

| | *Tax on transfer* | | |
Slice of cumulative chargeable transfers	*Cumulative total*	*Rate*	*Tax payable*
£	£	percent	£
< 128 000	0–128 000	nil	nil
> 128 000		20	20% of amount > 128 000

| | *Grossing-up of transfer* |
Net transfer	*Tax payable*
£	£
< 128 000	nil
> 128 000	nil + 1/4 of amount > 128 000

calculated on the value transferred *less* the tax. Hence, where the tax is paid by the donor then the amount of the tax becomes a further diminution in the value of the donor's estate and must be included in the transfer of value. Therefore a net-of-tax transfer made by the donor has to be 'grossed up' to arrive at the actual amount of the chargeable transfer (see Table 6.1 and Example 6.1).

It is necessary to gross-up lifetime transfers, but there is usually no need to gross-up transfers on death since the distributions from the estate will normally be made after inheritance tax has been paid.

Example 6.1

Assume that X makes a chargeable lifetime transfer of £200 000 to a beneficiary Y, having made no previous transfer and ignoring any reliefs.

(i) Transferor pays the tax

Tax payable on £200 000
First 128 000 @ nil	=	0
Next 72 000 @ 1/4	=	18 000 (see Table 6.1)
Tax payable by X	=	£18 000

(ii) Transferee pays the tax

Tax payable on £200 000
First 128 000 @ nil	=	0
Next 72 000 @ 20%	=	14 400 (see Table 6.1)
Tax payable by Y	=	£14 400

TAPERING RELIEF (FA 1986 Sch.19)

Lifetime transfers are subject to inheritance tax at half the full rate. The tax is adjusted to full rate if death occurs. However, the legislation permits a tapering scale of relief to be applied to transfers occurring within seven years of death. The tax is increased to full rate if death occurs within three years but for transfers which were made from four to seven years from the date of death then lower charges are levied as shown in Table 6.2.

Tapering relief is subject to a *de minimis* rule. If the tax calculated by applying tapering relief is less than the tax calculated by applying half death rates then the tapering relief will not apply. That is the tax payable cannot be reduced below that which would have been due if the transferor had not died (ITA 1984 s.7).

Table 6.2. Tapering relief

Years from transfer to death	Percentage of full tax rate	Actual rate
	percent	percent
< 3	100	40
> 3 but < 4	80	32
> 4 but < 5	60	24
> 5 but < 6	40	16
> 6 but < 7	20	8
> 7	exempt	—

INDEXATION (ITA 1984 s.8)

The rate bands (shown in Table 6.1) are indexed for inflation on an annual basis. On 6 April each year the rate bands are increased by the same percentage as the percentage increase in inflation. The increase in inflation is determined by comparing the Retail Price Index (RPI) for the preceding December with the RPI for the December twelve months before.

Adjustment to the rate bands for indexation can only be in an upward direction, and the resultant figure must be rounded to the nearest £1000. This annual adjustment of the rate bands is carried out by the issue of Statutory Instruments using figures produced by the Treasury.

EXCLUDED PROPERTY (ITA 1984 s.6)

Certain property is excluded from charge to inheritance tax. Excluded property is therefore left out of the value of an estate for inheritance tax purposes. Where a disposition occurs, either during lifetime or on death, the value transferred by a transfer of value will not take account of any excluded property.

It is clearly important to know what property is classified as excluded property, and a brief description of the types of excluded property is given below.

1. *Overseas property.* Property which is outside the United Kingdom is excluded, but only if the person with the beneficial interest is legally domiciled outside the United Kingdom.

   The legislation does not make specific provision for determining where property is situated and the general law of situs therefore applies. Whereas this may give complications for assets such as, say, bearer bonds, the position with landed property is clearer because situs depends upon actual physical location, about which there will normally be little doubt.

2. *Reversionary interests (ITA 1984, s.47)*. A reversionary interest refers specifically to a future interest accruing under a settlement. This may include an interest in a reversion accruing upon termination of an interest in possession (ITA 1984 s.50). Similarly, in Scotland, it may refer to an interest in the feu of a property subject to a life-rent (ITA 1984 s.47).

A reversionary interest will be excluded for the purpose of inheritance tax, unless one of the following applies.

(a) The reversionary interest had been bought.
(b) The interest is expectant on the determination of a lease which is treated as a settlement (because the lease is for life and was not granted for full consideration).
(c) The interest is a settlement which has been made after 15 April 1976 to which the settlor or that settlor's spouse is a beneficiary.

3. *Occupational schemes (ITA 1984, ss.58 and 151, F(No.2)A 1987, s.98(4))*. Where a scheme approved by the Superannuation Funds Office (IR) is offered by an employer to provide a retirement pension scheme, then this will be excluded.
4. *Government Securities*. Certain Government securities issued on terms giving exemption from taxation to non-residents who are not domiciled in the United Kingdom.
5. *Government pensions*. Certain pensions receivable from overseas countries (former colonies).[3]
6. *Savings of persons domiciled in the Channel Islands or the Isle of Man*. Persons domiciled in these locations and who are beneficially entitled to savings such as war loans, national savings and premium savings bonds can treat these items as excluded property.
7. *Visiting Forces*. Property in the United Kingdom belonging to members of Allied Military forces are excluded.

EXEMPT TRANSFERS (ITA 1984 ss.18–27)

A most important group of transfers is classified as being exempt from inheritance tax. These exempt transfers can be categorized with reference as to whether they apply only during the life of the taxpayer or whether they apply both during the life and on death. An additional category is that of general exemption.

General Exemptions
1. *Tax rate bands*. Clearly the nil rate of tax for the first £128 000 of chargeable transfer is itself an exemption.
2. *Gifts to charities (ITA 1984 s.23(1))*. Gifts to charities[4] made after 14

March 1983 are exempt without any limit. (Gifts made before that date were exempt up to a limit, latterly £250 000.)

If the value transferred is greater in the hands of the donor than in the hands of the charity the exemption will be in respect of the former, since it will refer to the loss to the tax payer's estate as a consequence of the disposition.

3. *Gifts to political parties (FA 1988 s.137)*. Gifts to political parties made after 14 March 1988 are exempt without any limit. (Gifts made before that date were exempt up to a limit.)

A political party will qualify for this exemption if it can demonstrate that at the General Election preceeding the transfer of value no less than two of its members were elected. If only one member was elected to Parliament the party may still qualify for these purposes if at least 150 000 votes were given to candidates who were its members.

4. *Gifts for national purposes (ITA 1984, s.25(1), Sch.3)*. This is a potentially important exemption for owners of land and property. The legislation permits exemption for transfer of value to certain national bodies including the following.

The National Gallery.
The British Museum.
The National Museum of Scotland.
The National Museum of Wales.
The Ulster Museum.
The National Trust.
The National Trust for Scotland.
The Historic Buildings and Monuments Commission for England.
The National Art Collections Fund.
The Trustees of the National Heritage Memorial Fund.
The Friends of the National Libraries.
The Historic Churches Preservation Trust.
The Nature Conservancy Council.
Any Local Authority.
Any government department.
Any university or college.
Any institution which exists for the purpose of preserving for the public benefit a collection of scientific, historic or artistic interest.
Any museum or art gallery maintained by a local authority or university.
Any library the main function of which is to serve the needs for teaching and research at a university.

This exemption is extended to sales of property to national bodies. The price which the buyer pays for a work of art or a building will take into account the exemption and the benefit of the exemption will accrue to both parties (ITA 1984 s.32 and FA 1985 s.26 Sch.4). For works of art etc. the transfer will receive 25% of the benefit of the tax exemption, while in the case of national heritage property the benefit will be limited to 10%.[5]

5. *Gifts for public benefit (ITA 1984 s.26 and FA 1985 s.95)*. Where a property is deemed to be eligible by the Board of Inland Revenue, and when that property is gifted to a body not established for profit, then the transfer of value may be exempt from inheritance tax.

This provision is again important for those with an interest in land and buildings, since property deemed to be ineligible by the Board may include lands of outstanding scenic, historic or scientific interest or buildings of outstanding historic, architectural or aesthetic interest where it may be necessary to find means of preserving them. Eligible gifts may include property as a source of income for the upkeep of property having the above characteristics.

Other eligible items may include pictures, prints, books, manuscripts, works of art or scientific collections.

When considering the eligibility of property for exemption in this regard the Board will want to be satisfied on a number of points including the following.

(a) Whether the receiving body is an appropriate one which will be effective in preserving the building or the character of the land.
(b) If property is gifted as a source of income for the upkeep of property in this category, then consideration will be given as to whether such income will be a sufficient amount.
(c) Whether undertakings may need to be sought in order to ensure adequate public access.

6. *Transfers between husband and wife (ITA 1984 s.18)*. Transfers made between husband and wife either during life, or on death, are exempt from inheritance tax.[6] Because of the absolute nature of this exemption transfers between spouses are not even classified as 'potentially exempt transfers'.

Exemptions on lifetime transfers only
These are certain transfers of value which can only be exempt if made by the taxpayer during his lifetime.

1. *Annual gifts*. There are two categories of annual gift.

(a) Transfers up to a value of £3000 (ITA 1984, s.19). Transfers of value in any given fiscal year, and during the donors lifetime may be made up to a limit (1988/89) of £3000.

If the full amount of the allowance is not transferred in one year the remaining amount can be carried forward, but not beyond the immediately following year (See examples 6.3 and 6.4 infra).

(b) Small gifts to the same person. Transfers of value to any one person up to an annual limit of £250 are exempt from inheritance tax provided that they are made during the lifetime of the donor.

This exemption applies to any number of gifts up to £250 to different individuals. It is in addition to the exemption of transfers up to a value of £3000, but the two exemptions cannot be combined to afford a larger gift to an individual.

In the case of both types of annual gift described in (a) and (b) above the exemptions will apply to both husband and wife. Also the value of the gifts will be calculated without reference to tax, i.e., they will not be grossed-up.

2. *Gifts in consideration of marriage.* Gifts made on the occasion of any one marriage by one donor will be exempt from inheritance tax. The value of the gift will be calculated without reference to tax.

Exemption from inheritance tax for such gifts are limited by reference to the familial relationship between the parties, on the following basis.

(a) £5000 if the donor is a parent of either marriage partner.
(b) £2500 if the donor is one of the marriage partners.
(c) £2500 if the donor is an ancestor of either marriage partner.
(d) £1000 if the donor is not classified in (a) or (c) above.

3. *Normal expenditure out of income.* This exemption is designed to cover the situation where there is a clear pattern of habitual and therefore normal expenditure which might otherwise be charged to inheritance tax.

In order to establish the bona fides for such an exemption the following conditions have to be met.

(a) It has to be shown that the transferor made the payment as a normal item of expenditure and that there is therefore an element of regularity (e.g., life assurance premium).[7]
(b) The payment has to be made out of income (not capital).
(c) It has to be shown that having transferred such normal expenditure out of income that the transferor is left with a sufficiency of income to maintain his standard of living at a level which would be considered usual in the transferor's circumstances.

4. Disposition for maintenance of family. A disposition during lifetime of a transfer of value will be exempt in a variety of circumstances relating to family maintenance, and including the following.

(a) A disposition from one party to a marriage to another in respect of maintenance.

(b) A disposition from one party to a marriage to another for the maintenance, education or training of a child up to the age of eighteen.

(c) A disposition designed to provide care and maintenance at a reasonable level for a dependent relative.

5. *Allowable transfers in the course of trade.* A disposition will not be considered a transfer of value if it is allowed for the purpose of computing profits for income tax purposes.

Exemptions for transfers on death only

1. *Death on active service (ITA 1984, s.154).* Where a person's death is attributable to a wound, accident or disease suffered or contracted while on active service against an enemy as a member of the armed forces then no inheritance tax is payable.[8]

   A leading case in respect of this exemption (which obtained for Capital Transfer Tax and Estates Duty) is that of Fourth Duke of Westminster's Executors v. Ministry of Defence[9] which established that the wound suffered in active service does not have to be the direct cause of death or the only cause of death. It will be sufficient that the wound was one of the contributory causes of death.

2. *Non-resident's bank accounts (ITA 1984 s.157).* When a person who is not domiciled in the United Kingdom dies then the balance held by that person in a qualifying foreign currency account is not included in the estate to be valued immediately before death.

   A 'qualifying foreign currency account' means any foreign currency account held with the Post Office, the Bank of England or any other bank recognized by the Banking Act, 1929.

3. *Quick succession relief.* Where a person, at the date of death, received property which was taxed as a chargeable transfer in the previous five years then his estate is entitled to relief.

   The means of providing relief is to reduce the tax charged on the person's estate by the percentages shown in Table 6.3.

Table 6.3. Quick succession relief

Time between transfers	Tax relief (%)
< 1 year	100
> 1 year but < 2 years	80
> 2 years but < 3 years	60
> 3 years but < 4 years	40
> 4 years but < 5 years	20

RELIEFS

Certain reliefs are afforded in particular sets of circumstances. These reliefs typically have a specific purpose and they often have direct implications for those with an interest in land and buildings.

1. *Business Property (ITA 1984 ss.103–114, FA 1986 ss.105 and 106, and Sch.19 and FA 1987 s.58, Sch.8).* Provided that certain conditions are satisfied then 'relevant business property' will qualify for relief from Inheritance Tax. The relief affords a percentage reduction in the value transferred and applies to transfers made during the lifetime of the transferor or on death.

   The conditions requiring to be met for business property relief are as follows.

   (a) The business must be a qualifying business ITA 1984 s.103(3). A 'qualifying business' must be a business carried on in the pursuit of a vocation or profession and for gain.

   (b) The property must be relevant business property. The kinds of business property qualifying for relief on the basis of their relevance will include the following.

       (i) Unincorporated business, i.e., a business or interest in a business such as sole trader or a partnership.

       (ii) The ownership of shares in a quoted or unquoted company where the shareholder has control (ITA 1984, s.269).

       (iii) Land buildings and/or plant provided they were deployed for use in a business carried on by a company controlled by the transferor or by a partnership of which the transferor was a partner.

          For any land, buildings or machinery owned by the transferor to qualify for relief the business which uses them (or the transferor's interest in it) must be relevant business property ITA 1984 s.105(6).

       (iv) Minority holdings of unquoted shares. Minority shareholdings are classified as either 'substantial minority holdings' – giving the owner 25% or more of the voting rights in the company; or 'other minority holdings'.

   (c) The property must have been owned for a minimum period. The property will not qualify for business property relief unless one of the following applies.

       (i) It was owned by the transferor for at least two years before the date of the transfer.

       (ii) It is property which replaced previous property which qualified, and both properties were held for at least two years during a five year period terminating with the transfer.

(d) Business property relief will only be available on a potentially exempt transfer where further conditions apply.

    (i) Where the transfer is during lifetime, then it is necessary to show that the property was owned by the donor throughout the period from the original date of transfer to the date of the donor's death.

    (ii) Where the transfer is on death it is necessary to show that the property was relevant business property.

The actual relief available to the taxpayer for property qualifying for business relief is a percentage reduction of the value transferred by the transfer of value.

The percentage rates applicable to different classes of relevant property are as follows.

- Unincorporated businesses   50%
- Shares giving control of a company   50%
- Land, buildings or machinery used by the taxpayer in a controlled company or a partnership   30%
- Unquoted minority shares – substantial minority holding   50%
- Unquoted minority shares – other minority holding   30%

The percentage reduction is made before the annual exemption and any other exempt transfers are deducted. The calculation is also made before grossing-up is carried out in circumstances where the transferor pays the tax (see Examples 6.6 and 6.7 infra).

2. *Agricultural Property (ITA 1984 ss.115–124 and FA 1986 s.105, Sch.19).* Prior to 1981 this important relief was available only to 'working farmers' who qualified only if they met specific conditions designed to ensure that they were mainly or wholly engaged in the practice of farming. Upper limits were also imposed on the acreage or value of the agricultural property qualifying for the exemption. Since the relief applies to farm houses and farm buildings as well as to farming land it is an important one which has been relaxed in terms of the type of agricultural property which will qualify and of the conditions which must be met by the transferor.

The conditions for relief are that the agricultural property must have been *occupied* by the transferor ~~for the two years~~ preceding the date of transfer, and for an agricultural purpose. Alternatively, it may have been ~~*owned* by the transferor for the seven years preceding~~ _{SUBJECT TO AN AGRIC. TENANCY SINCE} BEFORE ~~the date of transfer *and* have been~~ *occupied* during the period for an 10/3/1981 ~~agricultural purpose.~~ BY ANYONE

For the purpose of these conditions 'agricultural property' means agricultural land and woodland together with any building ancillary to such operations (e.g., for the intensive rearing of livestock). It also

ie Changes in blue = after 9/3/1992

includes farm houses and cottages as well as farm buildings. Farm houses can include such land together with the house as is appropriate to the style and location of the buildings.

'Agricultural property' now also includes land and buildings used for breeding and rearing horses.

The relief is a percentage reduction of the value transferred by a transfer of value. Currently the relief is 50% if the transferor enjoys the right to vacant possession, (or if he can obtain it within twelve months of the date of transfer). The relief is limited to 30% if this condition cannot be met, i.e., in situations where the agricultural property is let.

Agricultural value of agricultural property is based on the assumption that the property is held in perpetuity but subject to a covenant prohibiting its use for other than agricultural purposes.

Where farm cottages form part of the agricultural property and are occupied by persons employed exclusively in agriculture then they will be valued ignoring the fact that they would fetch higher prices if sold as residences to persons not employed in agriculture.

If relief is given for agricultural property then relief cannot also be claimed on the basis that it is business property. However, property not attracting agricultural relief may attract business relief with the convention that agricultural relief takes priority over business relief (see Example 6.7 infra).

CALCULATION OF THE TAX

Inheritance Tax is levied on the value transferred by a chargeable transfer. Added to this concept are the two principles of aggregation and cumulation. The result of the application of these principles is that the tax is levied on the value transferred by a chargeable transfer, at the rates obtaining at the time of transfer, to the highest part of the aggregate value of that transfer added together with all chargeable transfers made within a period of seven years from the date of the last transfer. Hence, any value transferred by chargeable transfer made more than seven years before the date of the last transfer is omitted.[10] When a transfer ceases to be cumulated it is removed as an item for which a calculation of tax payable has to be made (i.e., there is no repayment of tax).

In circumstances where one chargeable transfer includes more than one piece of property the tax chargeable on the aggregate value is apportioned between the assets according to the proportion which each bears to the aggregate (ITA 1984 s.265). Clearly this can be important where the incidence to tax may vary between assets because of, say, the incidence of reliefs.

Since the tax on the last transfer must take account of chargeable transfers previously made (and which remain chargeable) these may include ones made during the existence of Capital Transfer Tax which provided that as from 1981 chargeable transfers ceased to be cumulated if made more than ten years before the last chargeable transfer.

Example 6.2

Assume that A makes a chargeable transfer in June 1989 of £30 000. The history of previous chargeable transfers is as follows.

	£	
	£	
1. 1973	7 000	
2. 1976	25 000	(with the donee paying the tax)
3. 1983	72 000	
4. 1987	10 000	

Ignoring annual exemptions this liability to tax would be determined as follows.

Since Capital Transfer Tax was not introduced until 1974 the first gift is not chargeable.

The gift made in 1976 is chargeable. A's cumulative total will now be £25 000 and the charge to tax will be payable at the rates obtaining in that year.

The gift made in 1983 will require to be grossed-up since the donor is paying the tax. The value of the gross transfer will be calculated by applying 1983 tax rates to the cumulated value of £97 000 giving a new cumulative value of £102 975.

In 1986 the cumulative total will be reduced by £25 000 as the gift made in 1976 will cease to be included because of the ten year rule for transfers prior to 17 March 1986. The new cumulative total will be, therefore, £77 975.

The gift made in 1987 would be taxed at the then-obtaining rates and on the basis of a cumulative total of £87 975.

The gift made in 1989 would be taxed at the current rates and on the basis of a cumulative total of £117 975.

In 1990 the value transferred in 1983 would cease to be cumulated.

There are a number of additional points which can be made with reference to the calculation of tax.

1. Exempt transfers are not accumulated.[11]
2. Conditionally exempt transfers are not cumulated. Such transfers are those of qualifying 'national heritage property' (supra).
3. Potentially exempt transfers are not cumulated unless the donor dies within the seven year period (ITA 1984 s.3A(5) (1). It is assumed that a potentially exempt transfer is an exempt transfer until either the

seven years run out in which case it is exempt, or the transferor dies, in which case it becomes chargeable retrospectively. This will mean that transfer has to be taxed at the date at which it occurred (and not at the time of death). Clearly this will have an effect where a chargeable transfer occurs following a potentially exempt transfer. In these circumstances the chargeable transfer will have been assessed to tax on a cumulative value which did not include the potentially exempt transfer and this will have to be adjusted. The method for doing so will be to cumulate all chargeable transfers for seven years prior to the potentially exempt transfer (not the date of death). By this means a proper cumulative value can be determined for the taxation of the chargeable transfer. However, the process of reinstating potentially exempt transfers occurring within seven years of death as chargeable transfers is not wholly retroactive in effect since the rate of tax to be applied to such transfers will be that obtaining at the date of death.

The following examples illustrate the calculation of inheritance tax and include reference to land and building assets.

Example 6.3

A has made the following gifts.

1. 1 June 1987   £25 000
2. 1 September 1988   £100 000
3. 1 December 1989   £75 000

His position following the last chargeable transfer, and assuming that he has not made earlier transfers and that he bears the tax, will be as follows.

	Chargeable transfer 1 June 1987	25 000
Less	Annual Exemption 1986/87	3 000
	Annual Exemption 1987/88	3 000
	Cumulative net transfers	19 000
	Tax on gift (cumulative total < nil rate band limit of £90 000 for 1987–88)	0
	Chargeable transfer 1 September 1988	100 000
Less	Annual Exemption 1988/89	3 000
		97 000
	Cumulative net transfers	116 000
	Tax on gift 0–110 000	0
	110 000–116 000 (£6 000 × 1/4)	1 500

		Tax payable	1 500
	Chargeable transfer 1 December 1989		75 000
Less	Annual Exemption 1989/90		3 000
			72 000

Tax on gift	0–118 000	0
	118 000–188 000 (£70 000 × 1/4)	17 500
	Tax due	17 500
	Tax already paid	1 500
	Tax payable	16 000

		Gross	Tax	Net
	Cumulative totals	117 500	1 500	116 000
add	this transfer	88 000	16 000	72 000
		205 500	17 500	188 000

Example 6.4

Assuming the same circumstances as in Example 6.3 above but with the donee paying the tax then his position can be calculated as follows.

	Chargeable transfer 1 June 1987	25 000
Less	Annual exemption 1986/87	3 000
	Annual exemption 1987/88	3 000
	Cumulative net transfers	19 000
	Tax on gift (cumulative total <£90 000 for 1987/88	0
	Chargeable transfer 1 September 1988	100 000
Less	Annual Exemption 1988/89	3 000
		97 000
	Cumulative net transfers	116 000
	Tax on gift 0–110 000	0
	110 000–116 000 (£6000 @ 20%)	1 200
	Tax payable	1 200
	Chargeable transfer 1 December 1989	75 000
Less	Annual exemption 1989/90	3 000
		72 000
	Cumulative net transfer	188 000
	Tax on gift 0–118 000	0
	118 000–188 000 (£70 000 @ 20%)	14 500

Example 6.5

Assume that A dies on the 9 May 1990 leaving an estate with a total value of £225 000. The estate includes a 50% share of a house with a value of £100 000. The other 50% is owned by his wife. He bequeaths his 50% interest in the house to his wife together with half of the residue of his estate. The other half of the residue is bequeathed equally to his son and daughter.

The lifetime chargeable transfers which he had incurred during the seven years prior to his death totalled £49 000. Calculate the tax due on his estate.

Cumulative transfers prior to death			49 000
Value of estate at the date of death		225 000	
less exempt transfer (gift to wife) (50% of house)		50 000	
(50% residue of estate)		87 500	
Chargeable residue		87 500	87 500
Cumulative chargeable transfers			136 500
Tax due on death, i.e., on £87 500			
49 000–128 000 (£79 000 @ 0%)			0
128 000–136 500 (£18 500 @ 40%)			3 400
Tax payable			3 400

Example 6.6

In March 1990 A, who has made no previous transfers, settles the total value of his business jointly on his son and daughter. He has owned and managed the business since 1969. The business comprises land and buildings valued at £150 000, stock etc. valued at £50 000 and goodwill valued at £50 000 giving a total value of £250 000. The business has debts of £50 000.

From this information we can determine that the conditions necessary for the attraction of business property relief can be met. The business is a 'qualifying business' (a vocation or profession undertaken for gain). The property is 'relevant business property' (a sole trader deploying land and buildings for use in the business) and the property has been owned for the minimum period (two years).

The value of the business is the value of the assets of the business, (including the land and buildings and also including the goodwill), but less any liabilities attributable to the business (ITA 1984 s.227(7)).

Hence the value of the business transferred will be as follows.

		£	£
Value of unincorporated business:	stock	50 000	
	goodwill	50 000	
		100 000	
	less liabilities	50 000	
		50 000 less 50%	25 000
Value of land and buildings		150 000 less 30%	100 000
Total value of business (incl. land and buildings)		200 000	
Value transferred by the transfer of value (after relief)			125 000

Example 6.7

Mr F, a qualifying farmer, owns a farming business comprising a mixed arable and livestock farm extending to three hundred acres. The total value of the agricultural property included (on which relief is due) is £600 000. The total value of the business is £1 000 000. Assuming that the non-agricultural property satisfied the necessary conditions for business relief then the application of the reliefs and the determination of the value transferred would be as follows.

		£	£
	Total value of farming business		1 000 000
less	Agricultural relief on £600 000 @ 50%	300 000	
less	Business relief on £400 000 @ 50%	200 000	500 000
	Value transferred by the transfer of value (after relief)		500 000

Example 6.8

F transfers property valued at £250 000, part is non-agricultural and the 'agricultural value' of the agricultural property transferred is £200 000. A mortgage of £60 000 is secured on the whole of the property being transferred. The calculation of the value transferred for inheritance tax purposes will be as follows.

		£	£
	Value of agricultural property	200 000	
less	Proportion of mortgage applicable		
	$\dfrac{(200\,000 \times 60\,000)}{250\,000}$	48 000	
		152 000	
less	Agricultural relief @ 50%	76 000	76 000

	Value of non-agricultural property	50 000	
s	Balance of mortgage (60 000–48 000)	12 000	38 000
	Value transferred by transfer of value (after relief)		114 000

NOTES: CHAPTER SIX – INHERITANCE TAX

1. Capital transfer tax was introduced by the Finance Act 1975 ss.19–52 Schs.4–11.
2. Previous rates of tax were much less favourable and much more complicated than that currently obtaining (see Appendix 1). Note that extra tax resulting from death after 18 March 1986 due to chargeable transfers made before that date shall not exceed the tax that would have been payable under the old provisions.
3. These refer to funds receivable from funds set up under the Government of India Act 1935 s.273 or under the Overseas Pension Act 1973 s.2.
4. Charities are defined in ICTA 1988 s.506(1).
5. Further details of these arrangements can be got from IR67 'Capital Taxation and National Heritage'.
6. Such transfers are also exempt from capital gains tax.
7. Life assurance premiums will not qualify, however, if they are made out on an annuity purchased on the transferor's life.
8. For example, this exemption applies to the estates of those members of the armed forces who were killed on active service in the Falklands. The exemption also applies to members of the Royal Ulster Constabulary who are killed as a consequence of terrorist activity in Northern Ireland.
9. [1978] Q.B. CTTL.
10. Note that it is the 'value transferred' which is removed from cumulation, not the value of the gift itself. The 'value transferred' will be the net amount after making adjustments for the annual exemption and, if applicable, grossing-up.
11. There are special rules for allocating tax where the transfer is only partially exempt (see ITA 1984 ss.36–42).

Chapter Seven

CAPITAL GAINS TAX

INTRODUCTION

Short-term Capital Gains Tax was introduced by the Finance Act 1962 and amended by the Finance Act 1965. The Finance Act 1971 abolished short-term capital gains, and all capital gains are now dealt with under the long-term capital gains provisions of the Finance Act 1965 and subsequent legislation, principally including the Capital Gains Tax Act 1979 (CGTA 1979). The base year was changed from 1965 to 1982 by the Finance Act 1988.

Capital Gains Tax is levied on chargeable gains realized during the year of tax assessment (CGTA 1979 s.1). It is levied on gains accruing on or after 6 April 1965 on the disposal of all forms of asset including heritable, leasehold and other interests in property. However, for disposal made on or after 6 April 1988 Capital Gains Tax is only levied on gains accruing since the 31 March 1982 (FA 1988 s.96) but subject to certain restrictions.

Although Capital Gains Tax is separate from Income Tax, the procedures for returning information to inspectors, assessment and appeals are the same as for Income Tax.

With the harmonization of rates of Income Tax and Capital Gains Tax (in the financial year 1988–89) the differences in substance and effect have largely disappeared, and tax planning-measures are less crucial to tax minimization than has previously been the case during the whole of the life of Capital Gains Tax.

Nevertheless the two classes of taxation remain quite separate, and different reliefs and allowances apply. Should a gain be considered to be liable to Income Tax then it is excluded from assessment under Capital Gains Tax. Equally, losses which may be set-off against Income Tax may not be set-off against Capital Gains Tax.

OCCASIONS OF CHARGE

Capital gains tax is chargeable on the disposal of assets. For this purpose the disposal of an asset includes any occasion when the ownership of the asset is transferred, whether in whole or in part, from one person to another (except on death), for example by sale, exchange or gift[1], or when the owner of the asset derives a capital sum from it.

The tax is also charged on gains accruing on certain notional disposals, for example on gains which are deemed to accrue to trustees on the termination of a trust other than on a death of a life tenant.

Losses are treated in the same way as gains and are calculated accordingly on the basis, that if a disposal can result in a chargeable gain, then it should equally be capable of resulting in an allowable loss.

DISPOSALS

The concept of disposal is not defined in detail in the legislation. Disposal therefore has to be given its natural meaning. Accordingly an asset is disposed of whenever its ownership changes or whenever the owner of it divests himself of his rights in or interests over that asset.

The date of disposal and acquisition is statutorily defined as being the date of the contract, or the date of the conveyance (CGTA 1979 s.24). Where land is acquired under compulsory purchase powers the date of disposal and acquisition is the date when the compensation is agreed or determined, or, if earlier the date when the acquiring authority enters on to the land (CGTA 1979 ss.110 and 111). Disposal in respect of property will generally be by way of a sale or grant of a lease subject to a premium.

The concept of disposal is extended in three ways.

1. Part disposal. A part disposal arises when, on a disposal being made, any part of the asset remains (CGTA 1979 s.19(2)). Hence, if a fraction of a property is sold, then clearly there is a part disposal. However, the concept of part disposal can be carried through to a situation where an interest or right lesser than the original interest or right is created. For example, a sum derived from freehold property for the grant of a lease will be a part disposal.[2]

   Where there is a part disposal the proportion of the acquisition cost attributable to the part disposed of is A/(A + B) (CGTA 1979 s.35). Here, A represents the amount received for the part disposed of, and B is the value of the part retained (see Example 7.3).
2. Capital sums derived from assets. There is a disposal of an asset (in whole or in part) whenever a capital sum is derived from that asset, of which the following are examples.

(a) Capital sums derived by way of compensation for any kind of injury to assets, (such as, say, injurious affection suffered by land).[3]
(b) Capital sums received under a policy of insurance.
(c) Capital sums received in return for the surrender of rights (such as the release of a restrictive covenant).
(d) Capital sums received as a consideration for use or exploitation of assets, (such as the receipt of a premium for a lease over land).

3. Settled property. Trustees of settled property are (in certain circumstances) treated as disposing of assets both where the asset ceases to be settled property and where it remains in the settlement.[4]

Allowable expenditure
When calculating the incidence of gain in the disposal of an asset there is provision for the deduction of specific items in order to arrive at the actual gain (CGTA 1979 s.42).

These items are as follows.

1. The cost of acquisition (or the value of the asset when acquired).
2. The costs of acquisition (fees, valuations etc.).
3. Any expenditure on improvements which have increased the value of the asset.
4. Any costs incurred in protecting rights in the asset.
5. The costs of disposal (agents commission etc.).

(See Example 7.6)

RATE OF TAX

From the date of its original introduction until the financial year 1987/88 Capital Gains Tax was applied at a flat rate of 30%. However, from 1988/89 chargeable gains are taxed at Income Tax rates. In the case of individuals therefore net chargeable gains in any year of assessment will be taxed at the top slice rate applicable to that person's Income Tax (FA 1988 s.98). Currently companies liable to assessment for Corporation Tax will be charged to Capital Gains Tax at the full Corporation Tax rate of 35%.

Individuals may claim an annual exemption. For 1990/91 the exemption is in the sum of £5000, but this amount is index-lined (FA 1982 s.80).

PERSONS LIABLE TO TAX

All persons and companies ordinarily resident in the UK are liable to pay Capital Gains Tax on chargeable gains accruing in any year of assessment (CGTA 1979 s.2). Individual partners, and not the partnership, are responsible for tax chargeable on gains in so far as the partner gains even although the gains are realized by the partnership (CGTA 1979 s.60). Trustees are responsible for the payment of tax on gains accruing in the course of administration of the estate.

With the introduction independent taxation of married women from 6 April 1990 each partner is liable for capital gains tax on the disposal of their individual assets and at their individual rates of tax. Both husband and wife can claim an exemption from the first £5000 of gain in any year.

As we have seen in Chapter 3 a company which pays Corporation Tax has Capital Gains charged to Corporation Tax and not to Capital Gains Tax.

Exemption from Capital Gains Tax is offered to local authorities, charities, approved superannuation funds, friendly societies and registered trade unions.

INDEXATION

Until 1982 gains were calculated on the basis that they were due on the difference in value in actual prices between acquisition and disposal but on the assumption that no gains (or losses) accrued prior to 6 April 1965. (The taxpayer could elect to use the actual value as at 6 April 1965 for an asset acquired before that date or the figure that would have obtained on that date calculated on a basis of straight line time apportionment.)

The effect of this method was to include inflationary as well as real gains in the value of assets in the calculation for tax. In recognition of this unfair effect an indexation allowance has been introduced since the financial year 1982–83 (FA 1982 ss.86 and 87 and Sch.13). Inflationary gains made between 1965 and 1982, however, remained to be taxed as before.

Since 1988 this anomaly has been dealt with by treating assets acquired before 31 March 1982 as having been disposed of upon that date and immediately re-acquired (at market value) (FA 1988 s.96). While introducing this provision the election to use time apportionment has been dispensed with (see Example 7.1).

While the rules incorporating a 1982 re-basing are now more equitable and factual they are not applied without restriction. If the current rules produce a higher gain (or greater loss) then the former rules will apply in their place.[6] Clearly this will require a calculation to be made on both sets of rules to determine the actual position. This in turn will require the preparation of valuations for historical dates or the application of time apportionment.

It would appear that in calculating any gain or loss under the former rules, then indexation should apply[7] (see Example 7.2). Each item of allowable expenditure will be treated separately. The aggregate of each calculation will be the indexation allowance.

Only allowable expenditure can be indexed (CGTA 1979 s.32). Relevant items will include the following.

1. The cost of acquisition.
2. The expenses of acquisition.
3. Money spent on improving the value of the asset.
4. Expenses incurred in defending the ownership or rights in the asset.
5. Disposal expenses.

They will not include, for example, inheritance tax paid on a gift the disposal of which also results in a capital gain.

To qualify for an indexation allowance the disposal must be after 6 April 1982. In addition the disposal must be more than twelve months after the acquisition of the asset.

Assets held on 6 April 1965

Although capital gains tax has been re-based to 31 March 1982 the rules pertaining to assets held on 6 April 1965 may still apply since re-basing only applies if the effect is to reduce a gain or a loss.

The taxpayer will have the right to choose, on the basis of comparative advantage, between the application of a time-apportioned valuation of the asset at 6 April 1965 or the actual market value at that date (see Example 7.4).

With assets which have been held since before April 1965, it can often be the case that additional capital has been spent on altering, improving or defending the interest. Where expenditure has been made which adds value to an asset then the contribution that such improvements make will be apportioned for the purpose of establishing the chargeable gain (see example 7.5).

Example 7.1

A purchased a block of land in March 1980 for £50 000. He sold the land in August 1987 for £125 000. The RPI for March 1982 (RI) = 79.4 and the RPI for August 1987 (RD) = 101.8.

The chargeable gain will be calculated as follows.

		£
	Amount of consideration for the asset	125 000
less	original acquisition cost	50 000
	Unindexed gain	75 000
Indexation allowance:		

$$\frac{RD - RI}{RI} = \frac{101.8 - 79.4}{79.4} = 0.282$$

	£
50,000 × 0.282	14 100
Chargeable gain	60 900

Example 7.2

A purchased a block of land in 1976 for £25 000. The market value of the land on 31 March 1982 was £55 000. In May 1988 A sold the land (with benefit of planning permission for housing) for £500 000.

The chargeable gain will be calculated as follows

		£	£
	Amount of consideration for the asset	500 000	500 000
less	original acquisition cost	25 000	
or			
less	market value as at 31 March 1982		55 000
	Unindexed gain	475 000	445 000
	Indexation allowance:		

$$\frac{RD - RI}{RI} = \frac{106.2 - 79.4}{79.4} = 0.338$$

	£	£
55 000 × 0.338	18 564	18 564
	456 436	426 436
Chargeable gain		426 436

Example 7.3

A purchases land and buildings in 1969 for £10 000. A firm of Chartered Surveyors valued the whole of the property at £80 000 in March 1982. He sold part of the land for residential development in April 1988 for the sum of £100 000. The surveyors estimated that the value of the land and buildings retained was £65 000.

The chargeable gain on the part disposal will be calculated as follows.

£

Amount of consideration on (part)
disposal

100 000

Apportioned acquisition cost:

$$\text{Original cost} \times \frac{A}{A + B}$$

$$10\,000 \times \frac{100\,000}{100\,000 + 65\,000} \quad = \quad 6061$$

Indexation allowance:

$$\frac{RD - RI}{RI} = \frac{105.8 - 79.4}{79.4} = 0.332$$

$$80\,000 \times \frac{100\,000}{100\,000 + 65\,000} \times 0.332 \quad = \quad 16\,096 \qquad 22\,158$$

Chargeable gain on part disposal
of asset

= 77 842

Example 7.4

A purchased land in April 1960 for £5000. He sold it in April 1989 for £70 000. In April 1965 a firm of Chartered Surveyors valued the land at £5800.

Ignoring indexation the chargeable gain may be calculated on the two alternative basis.

1. Straight line method

£

Amount of consideration received
for the asset

70 000

less original acquisition cost

5 000

Gain over the whole period

65 000

Adjustment of acquisition cost to
6 April 1965

$$\frac{\text{April 1989} - \text{April 1965}}{\text{April 1989} - \text{April 1960}} = \frac{24}{29} \times 65\,000 \qquad = 53\,793$$

Chargeable gain

= 53 793

2. April 1965 valuation

		£
Amount of consideration received for the asset		70 000
less value as at 6 April 1965		5 800
Chargeable gain		= 64 200

An election for an April 1965 valuation would not be made.

Note: See 'Assets held on 6 April 1965', supra.

Example 7.5

Assume the same factors as in Example 7.4 with the following added. A spent sums improving the assets as follows.

	£
April 1963	500
Sept. 1969	1000
Jan. 1976	2500

Ignoring indexation the chargeable gain may be calculated on the two alternative basis.

1. Straight line method

Amount of consideration received for the asset			70 000
Original acquisition cost		5000	
Cost of improvements	I_1	500	
	I_2	1000	
	I_3	2500	9 000
Gain			61 000

Apportionment of gain.

$$\text{Acquisition cost} = \frac{5000}{9000} \times 61\,000 \quad = 33\,889$$

$$I_1 = \frac{500}{9000} \times 61\,000 \quad = 3\,389$$

$$I_2 = \frac{1000}{9000} \times 61\,000 \quad = 6\,778$$

$$I_3 = \frac{2500}{9000} \times 61\,000 \quad = 16\,945$$

$$61\,000$$

Time apportionment

$$\text{Acquisition cost} \quad \frac{\text{April 1989 − April 1965}}{\text{April 1989 − April 1960}} = \frac{24}{29} \times 33\,839 = 28\,064$$

$$\text{I}_1 \quad \frac{\text{April 1989 − April 1965}}{\text{April 1989 − April 1963}} = \frac{24}{26} = 3\,389 = 3\,128$$

I_2 (post April 1965)	=	6\,778
I_3 (post April 1965)	=	16\,945
Chargeable gain		54\,915

2. April 1965 valuation

			£
Amount of consideration received for the asset			70\,000
less Value as at 6 April 1965		5800	
Expenditure on post 1965 improvements	1000		
	2500	3500	9\,400
Chargeable gain			61\,600

An election for an April 1965 valuation would not be made.

Note: Indexation has been ignored in this example, see 'Assets held on 6 April 1965', supra.

Example 7.6

A sells a shop property held for investment in May 1988 for £300 000. The shop had been acquired in April 1981 for a sum of £150 000. During 1981 A spent £5000 of capital expenditure on improvements. He also spent £5000 carrying out repairs to the fabric of the building and renewing external decoration. He incurred legal costs of £2500. Prior to purchase he instructed a firm of Chartered Surveyors to value the property and he received a fee account of £750.

At 31 March 1982 the market value of his interest in the property was £115 000.

In order to effect the sale in 1988 Mr A incurred advertising expenses of £500 and agent's fees of £6000. Legal costs amounted to £1000.

Making the assumption that Mr A's profit on the transaction will be chargeable to Capital Gains Tax, and that he has no other capital gains or losses to take into account, the chargeable gain will be calculated as follows.

			£
	Proceeds on disposal		300 000
Less:	advertising	500	
	agent's fees	6 000	
	legal fees	1 000	7 500
			292 000
add:	acquisition costs	150 000	
	expenditure on improvements	50 000	
	repairs and decoration	nil	
	legal fees	2 500	
	valuation fees	750	
	Total cost	203 250	
add:	Indexation		

$$\frac{RD - RI}{RI} = \frac{106.2 - 79.4}{79.4} = 0.338$$

115 000 × 0.338	38 816	
	242 006	242 066
Chargeable gain		49 934

ASSETS

Subject to a number of exemptions (some of which are detailed below) all forms of property will be treated as assets for the purpose of Capital Gains Tax.

Assets or property are not defined in the legislation although it has been established that the following categories of asset are property.

1. Investments.
2. Jewellery, antiques etc.
3. Land and buildings.
4. Options, debts and incorporeal property.
5. Any currency other than sterling.
6. Any form of property created by the person disposing of it.[8]

Considering property comprising land and buildings then any proprietary right in these will be assets for the purpose of Capital Gains Tax. Any estate or interest in land which comprises the heritable or leasehold interest will be a proprietary right and thus an asset. Such proprietary rights will be assets even though they may not be assignable.[9]

This leaves the position of certain rights in property unclear. Some rights in land are personal, such as the right of an individual to occupy premises (without having a right to an interest in land).

It is probably more prudent to work on the practical basis that if a gain accrues on the disposal of any asset then the Inland Revenue will see a *prima facie* case for chargeable gain to be taxed.

EXEMPTIONS

Certain classes of asset are neither chargeable gains nor allowable losses on disposal. These include the following.

1. Private motor vehicles (CGTA 1979, s.130)[10]
2. National Savings Certificates and other non-marketable securities issued under the National Loans Act 1939 (CGTA 1979, s.71).
3. Gambling winnings (including premium bonds) (CGTA 1979 s.19(4)).
4. Gains from foreign currency obtained for personal expenditure abroad.[11]
5. Gains resulting from disposals of a persons decorated for gallantry (CGTA 1979 s.131).
6. Rights accruing from life assurance policies and deferred annuities are not chargeable gains provided always that the tax payer was the original beneficial owner and the rights were not acquired from someone else (CGTA 1979 s.143).
7. A chargeable gain or allowable loss will not accrue on disposal of a debt provided always that the tax payer is the original creditor (CGTA 1979 s.134).
8. Wasting assets which include moveable property (not land and buildings) will not suffer chargeable gains on disposal (CGTA 1979 s.127(1)).[12]
9. Chattels disposed of for £6000 or less (CGTA 1979 s.128(1), FA 1989).
10. Disposal of a main private residence (CGTA 1979 ss.101–105).
11. Disposal of British government securities (FA 1986 s.59).
12. Gifts to the nation of works of art considered to be of national interest are exempt. This exemption also includes land and buildings given to the National Trust (CGTA 1979 s.147).
13. Gifts to charities (CGTA 1979 s.145(1)).
14. Disposal of certain corporate bonds issued after 13 March 1984 (FA 1985 s.67).
15. Gains or losses accruing from transactions in future and qualifying corporate bonds after 1 July 1986 (FA 1986 s.59).
16. Gains made on business expansion scheme shares issued after 18

March 1986 provided always that the tax payer is the original owner (TA 1988 ss.289–312 and FA 1988 ss.50–53 and Sch.4).

17. Where land is given or transferred cheaply to a registered housing association it will no longer be treated as taking place at market value so that a capital gain will only arise if actual proceeds exceed allowable expenditure. If they are less, the landowner will be treated as making neither a gain nor a loss after any indexation allowance.

MAIN PRIVATE RESIDENCE

Any gain accruing on the disposal of a property which has been the only or main residence of the tax payer is exempt from Capital Gains Tax (CGTA 1979, s.128(4)). It is not necessary that the house is occupied by the tax payer at the time of the disposal. It is sufficient that the house was the tax payer's principal residence at any time during the period of ownership.

Land may be included with the residence in the exemption provided that this is appropriate garden or amenity land which is for the exclusive occupation and enjoyment of the tax payer together with the dwelling. However, if the land accompanying the dwelling is in excess of one acre then the exemption will only apply if the Inland Revenue can be satisfied that the area of land is appropriate to the reasonable enjoyment of the dwelling having regard to the size and character of the house.

If it is the intention of the tax payer to realize a gain from the disposal, or if the tax payer extends or improves the property again with the specific purpose of making a gain then the exemption will be lost or there will be a charge on that part of the gain attributable to the additional expenditure (CGTA 1979 s.103(3)).

Since the majority of people in the community will own a house or be dependent upon such a person, and since a private residence will be the single most important asset of most taxpayers, then this exemption is clearly of some moment to a large section of the population.

However, for the exemption to be obtained several rules have to be satisfied, as follows.

1. The property must have been used as an owner-occupied residence. (Periods of residence and ownership prior to 6 April 1965 are ignored.)
2. The degree of exemption is proportionate to the extent to which, during the period of ownership, the house has been occupied by the owner as a principal private residence (CGTA 1979 s.102(2)).
3. Certain periods during which the owner was absent from the property are to be disregarded provided that both before and after these

periods the house was the owner's only or main residence and that throughout these periods he had no other house eligible for exemption.

These are as follows.

(a) The last twenty-four months of ownership[13] (CGTA 1979 s.102(1)).
(b) Any period (or periods) of absence not totalling more than three years.
(c) Any period during which the owner was working outside the UK.
(d) Any period(s) of absence not totalling more than four years throughout which the owner of the house was prevented from living in it because his employer required him to live elsewhere in order to do his job effectively (CGTA 1979 s.102(3)).
(e) Any period (from July 1978) during which the owner had to live in tied accommodation but with the intention of returning to the principal residence.

Example 7.7

A bought a house on 1 January 1974 for £20 000 and resided in the house until 1 June 1980. On the 1 June 1989 he sold the house for £120 000. He had no other residence either when he occupied the house or thereafter. Assuming that his absence from the house did not include qualifying periods then A will have been deemed to have occupied the house for 8.5 years of the total period of ownership 15.5 years. The 8.5 years will be made up of 6.5 years between 1 January 1974 and 1 June 1980 when he was in residence together with the last two years of his ownership of the house.

The chargeable gain will therefore be calculated as follows.

		£
	Amount of consideration received	120 000
less	original acquisition cost	20 000
		100 000
less	exemption for owner occupation	
	$\dfrac{6.5 + 2}{15} \times 100\,000$	56 667
	Chargeable gain	43 333

Note 1: Indexation has been ignored.

2: Consideration should be given as to whether a calculation made on a base of 31 March 1982 would give the taxpayer a more favourable outcome.

4. In a situation where the use of the dwelling is divided between residential and business use then the proportion of the gain realized attributable to the business use will be non-exempt[14] (CGTA 1979 s.103(1)).

5. In a situation where part of the tax payer's main residence is let, then, since 6 April 1980, the exemption can extend to the tenanted part. However, the amount of the exemption on the let part cannot exceed the amount of the exemption on the owner-occupied part, or £20 000, whichever is the smaller[15] (FA 1984, s.63). Note, however, that the whole of the house may be let during the allowable periods specified in 3, supra.

6. If a taxpayer owns two residences then an election may be made as to which residence is his 'main' residence for Capital Gains Tax purposes (CGTA 1979 s.101(5)). The election must be made within two years of the acquisition of the second residence. The default position is that the Inland Revenue will decide for themselves which dwelling is the main residence and this decision will be based on the facts of the case, particularly including the periods of time actually spent by the tax payer at each house.

7. Spouses (who are living together) can claim only one exemption from Capital Gains Tax in respect of the residence jointly occupied by them[16] (CGTA 1979, s.101(6)).

BUSINESS ASSETS

Replacement of business assets

The disposal of assets which are used in a business will give rise to chargeable gains or allowable losses for the purpose of Capital Gains Tax. However, if other assets are acquired to replace those which have been disposed of, then the owner may elect to defer liability to Capital Gains Tax by means of 'roll-over' relief (CGTA 1979, s.115). The effect of this relief is that any gain accruing on the disposal of the old assets may be deducted from the cost of the new assets.

Hence, no tax is paid until these new assets have been sold. But these new assets may in turn be replaced in time by other qualifying assets and tax payment again may be deferred. There may be other hidden benefits. Since the relief is actually a reduction of the cost of the new assets it is not simply a deferral of the tax payable on the gain made at that time. The new assets may be disposed of at a loss and the liability to tax will be expunged. If the new assets are eventually sold at a gain some other form of relief, such as retirement relief, may be available.

In order to qualify for the relief the new business assets must be acquired within a period commencing twelve months before the date of disposal of the old assets and ending three years after that date (CGTA 1979 s.115(3)) (see Example 7.8).

The new assets must be used in the same business as the old and the old assets must have been used throughout the period of ownership only in connection with the business.[17] If there was a qualifying business use for only part of the period then an apportionment must be made (CGTA 1979 s.115(6)).

Roll-over relief on the replacement of business assets is only available for certain types of asset specified in the legislation (CGTA 1979 ss.115(1) and 118). These include the following.

1. A building (or part of a building), or any permanent (or semi-permanent) structure in the nature of a building which is used for the purposes of a trade or occupation.
2. Land used or occupied for the purpose of a trade.[18]
3. Fixed plant and machinery, (not forming part of a building).
4. Shops, aircraft and hovercraft.
5. Goodwill.
6. Satellites, space stations, space vehicles and launch vehicles.[19]
7. Milk quotas and potato quotas.[20]

There is a modification to these rules where the new asset acquired under this relief is a wasting asset.[21] The modification is that in these circumstances the gain is held over instead of being deducted from the cost of the wasting asset. The period over which it will be held will be until the shorter of the following.

1. The new asset is disposed of.
2. The new asset ceases to be used in connection with the business.
3. The end of a period of ten years.

Example 7.8

A purchased a building in May 1976 for £25 000. He sold it for £39 000 in June 1980 and replaced it with another building which he acquired in July 1980 for £45 000. He sold this building in May 1990 for £65 000.

Taking into account the relief available for the replacement of business assets and ignoring indexation his chargeable gain can be calculated as follows.

	£
Financial year 1990–91	
Proceeds of disposal	39 000
Cost of acquisition	25 000
Gain	14 000

(No capital gains tax payable in this year because the business asset is replaced.)

Financial year 1990–1991

	Proceeds of disposal		65 000
Less	cost of acquisition	45 000	
	gain from 1980–81	14 000	31 000
	Chargeable gain		34 000

Business retirement relief

A tax payer who is over sixty years old makes a 'material' disposal of business or business assets may be eligible for business retirement relief[22] (CGT 1979 ss.124 and 125; FA 1985 ss.69 and 70). This special relief may also be available to a person who has retired at an age less than 60 because of ill health.[23]

As well as the tax payer being a qualified person for the purpose of this relief the nature of the disposal must also qualify. The legislative definition includes the following.

1. A disposal of whole or part of a business.
2. One or more assets which were in use in the business at the date upon which trade ceased (FA 1985 s.69(2)).

As to what is a qualifying business this will be a question determined with reference to the facts of the individual case. The courts have stated that management of property is not a business.[24] On the other hand the legislation specifies that retirement relief is available for furnished holiday lettings (FA 1984 Sch.11 para.1(2)(g)).

Clearly the definition permits the disposal of assets formerly used in a business after the business has ceased trading, for the purpose of this relief. The question of what is an asset used in the business and what is part of a business remains. In McGregor v Adcock it was held that the sale of a few acres from a larger farm did not materially affect the scale of the business and since the business was continuing no relief was available for the disposal of an asset.[25]

The assets disposed of must qualify as 'chargeable business assets' as defined in the legislation (FA 1985 Sch.20 para.12 (2)). The definition identifies a chargeable business asset as an asset (or an interest in one) used for the purposes of the trade carried out by a person (and including goodwill).[26]

If the qualifying person disposing of the business (by gift, sale; in whole or in part) has owned the business for the preceding ten years then there is an exemption from capital gains tax on the first £125 000 accruing from any gain on chargeable business assets.

If the disposal occurs after 5 April 1988 an additional exemption of 50% is given on the next slice of gain up to £375 000 after the first £125 000. Any excess over £500 000 is taxable in full (FA 1988 s.110).

The same level of relief is afforded on the disposal of shares in a trading 'family company', provided that the tax payer has been a full-time director in the company for the preceding ten years.[27] A spouse will also be eligible for the relief if the same conditions are met.

Where the qualifying conditions have been met but for a period less than ten years, then relief is afforded on the basis of a *pro rata* reduction in respect of the period held with 10% of the full relief being available for each complete year of ownership (see Example 7.9).

However, the length of the qualifying period may be increased if other businesses owned in the preceding ten years can be taken into account (FA 1985 Sch.20 para.14(3)).

Gifts of business assets

Where a tax payer disposes of business assets by means of gift or other than at market value than any chargeable gain accruing may be 'held-over'[28] (FA 1980 s.79).

The business assets to which this relief applies include the following.

1. An asset used in the tax payer's trade, profession or vocation.
2. An asset used in a family company (involved in trade).
3. Shares or securities in a family company (involved in trade) (CGTA 1979 s.126(1)).

The relief also specifically applies to commercially managed woodlands (CGTA 1979 s.126(8)), and to furnished holiday lettings (FA 1984 Sch.11 para.1(h)).

For the purpose of this relief it is necessary that both the donor and the donee jointly claim the relief (CGTA 1979 Sch.4 para.5).

The most recent provision for gifts of business assets is that on the disposal of business assets (acquired before March 1982) before 6 April 1988 when one half of the gain will be excluded.

GIFTS

A gift is a disposal for Capital Gains Tax purposes. Where a person acquires an asset by way of gift or by any other means than by way of a bargain made 'at arm's length' then the person is deemed to have acquired it at market value (CGTA 1979 s.29A). It may be in these circumstances that the donee has to pay tax on the benefit received and not recovered from the donor (CGTA 1979 s.59).

Example 7.9

A had a photographic business which he established on 1 September 1982. On the 31 May 1988, aged 61½ he sold the business to a large chain of photographic suppliers and processors.

The assets of the business are as follows.

		£	
	Goodwill	200 000	
	Premises	75 000	
	Equipment	7 000	(no capital
		282 000	allowances)
	Debts	5 000	

		£
Proceeds on disposal		277 000
Cost of acquisitions		
Premises	50 000	
Equipment	10 000	60 000
Unindexed gain		217 000
Indexation allowances		

$$\frac{RD - RI}{RI} = \frac{106.2 - 81.9}{81.9} = 0.297$$

$60\,000 \times 0.297$	17 820
Gain	199 180
Retirement relief	

$$\frac{5\%_{12}}{10} \times 125\,000 = 71\,875$$

$\frac{1}{2}(202\,300 - 71\,875) = 65\,212$	137 082
Chargeable gains on disposal	62 098

However, 'hold-over' relief will be available to transfers of assets other than bargains made 'at arm's length' after 5 April 1980 (CGTA s.126). Both the transferor and the transferee are required to claim such an election.

The effect of this relief is to reduce the capital gain to nought. The donee deducts the donor's capital gain from his cost of acquisition of the asset.

CHATTELS

There is special provision in capital gains tax legislation in respect of assets which are tangible moveable property where the value of the consideration on disposal does not exceed £3000 (CGTA 1979 s.128). Tangible moveable property refers to items such as works of art or furniture.

There is no chargeable gain on the disposal of such an asset and the consideration will be the gross amount (before any expenses of disposal).

If such an asset is sold for more than £3000 then the amount of capital gain is restricted to five-thirds of the difference between the actual consideration and £3000 (CGTA 1979 s.128(2)). Thus if an item of tangible moveable property is sold for £3600 then the capital gain is limited to $(3600 - 3000) \times 5/3 = £1000$.

WASTING ASSETS

Wasting assets are subject to special rules on disposal which effectively restrict the amount of allowable expenditure (CGTA 1979 s.127). A wasting asset is one which has a predictable life not exceeding fifty years (CGTA 1979 s.37 and s.127(5)).[29,30]

The wasting asset provisions will not apply to assets which have, or could have, capital allowances claimed against them.

The method of dealing with wasting assets for Capital Gains Tax purposes is to deduct the residual value from the cost of acquisition and to write off the amount arrived at over the life of the asset unexpired at the date of acquisition, on a straight line basis (CGTA 1979, s.38).

Since wasting chattels and assets which can attract capital allowances are excluded from the application of these rules, then there are limited instances where they will apply. Such instances will generally be in respect of intangible property such as options and life interests in settled property.

LEASES

The particular case of a lease of land for a term of fifty years or less is given special treatment in Capital Gains Tax legislation. Although it is a wasting asset the write-off provisions are not based on a straight line method but on a form of declining percentage.

Another distinctive feature of a 'short' lease for Capital Gains Tax purposes is that where a lease is granted subject to the payment of a premium then there will be a part disposal of the asset (CGTA 1979 Sch.3 para.2(1)). A premium is any sum paid to a landlord in consideration of the grant of the lease.[31]

Where a part disposal occurs then the usual part disposal rules will apply with the obvious difference that the market value of that part of the asset remaining will comprise the value of the rent accruing for the duration of the lease capitalized into a single sum (CGTA 1979 Sch.3, para.2(2)).

The method of writing off expenditure in a wasting lease of land is dictated by the application of percentage tables (CGTA 1979, Sch.3 para.1(3)). The effect is to provide curved line depreciation allowing for modest depreciation in the early years of the lease with rapidly accelerating depreciation as the end of the term is approached.

Statute provides, therefore, that the acquisition cost will be written off to the date of the disposal by the application of the fraction

$$\frac{P(1) - P(3)}{P(1)}$$

where, $P(1)$ = percentage applicable to unexpired term at the date of acquisition

and

$P(3)$ = percentage applicable to unexpired term at the date of disposal.

(See Example 7.10.)

If any monies have been expended on improving the interest then such expenditure will be modified by the application of the fraction

$$\frac{P(2) - P(3)}{P(2)}$$

where, $P(2)$ = percentage applicable at the date of the expenditure (or when the value of the expenditure was reflected in the lease).

Example 7.10

A acquired a lease of a shop on 1 January 1984 for £50 000. The lease had a term of ninety-nine years with an expiry date of 1 January 2007. On the 1 April 1988 A assigned the lease for a sum of £100 000.

The calculation to determine the amount of the lease which may be written off as a wasting asset and the amount chargeable to Capital Gains Tax is computed, with reference to the above facts, as follows.

		£
Proceeds on disposal		100 000
Cost of acquisition	50 000	

Fraction written of = $\dfrac{P(1) - (P(3)}{P(1)} = f$

$P(1) = 23$ years; % = 78.055
$P(3) = 19$ years; % = 70.791

$$f = \frac{78.055 - 70.791}{78.055} = 0.093$$

Amount of expenditure wasted = 50 000 × 0.093	4 653	45 347
Gain		54 653

Indexation allowance

$$\frac{RD - RI}{RI} = \frac{105.8 - 86.8}{86.8} = 0.219$$

54 653 × 0.219	11 963
Chargeable gain	42 690

The '3rd Schedule' provisions

Surveyors and valuers will be interested to note that the figures which appear in the 3rd Schedule table (see Appendix 2), and which fix the percentage of the original cost remaining in a given year are derived from Parry's Valuation Tables. The depreciation percentage is calculated by using the Year's Purchase Table (6%, single rate), so that

$$a = \frac{100}{15.7619} \times \left| \frac{1 - (1.06)^{-n}}{0.06} \right|$$

where,   a = depreciation percentage
0.06 = rate of interest
n = number of years
15.7619 = YP for 50 years at 6%

Effectively, the 3rd Schedule provision therefore allows for a curved-line depreciation in the value of the (leasehold) asset as against the

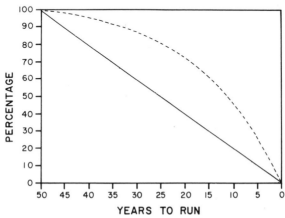

Figure 7.1: 'Straight-line' and 'curved-line' depreciation compared

straight-line depreciation allowed for all other wasting assets. The assumption underlying the method is correct since it anticipates that the capital value of a leasehold will be retained in its early years while experiencing accelerated depreciation towards the end of the term, rather than depreciating at an even annual rate throughout the whole of its life (see Fig. 7.1).

This general assumption is correct, although no reference will be found as to why a curve having these particular characteristics has been chosen.

The 3rd Schedule provisions had their origin in the Finance Act 1965 and (with the benefit of hindsight) it can be shown that the curve selected by the legislators very considerably underestimates the actual capital value of leaseholds over time.

Research has shown that the capital value profile of leases for fifty years or less rise for most of their life, and retain a value above their original cost for all but the last five years of the term, when they decay very rapidly indeed[32] (see Fig. 7.2 and 7.3).

It would appear on this evidence that (in aggregate) 'short' leases only depreciate over the last few years of their life, and then at an accelerated rate which is represented by a virtual straight line. This would suggest that short leaseholds should be amortized on a straight-line basis over the final five years of their lives even if growth rates are zero and rates of return are in modest single figures. Since government-projected growth rates and rates of return were both high in the 1960s it is clear that legislators were concerned to propose a curved-line of depreciation that

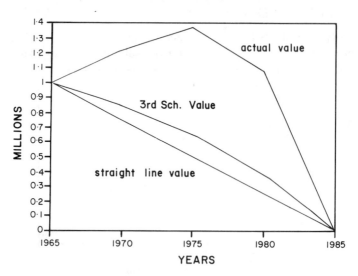

Figure 7.2:   Capital value of leaseholds, 1965–85

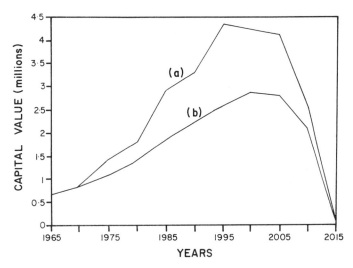

Figure 7.3: Actual and project value of leasehold investment over fifty year period
(a) Growth rate = 5% p a, IRR = 12% reducing to 4% at 5 year intervals.
(b) Growth rate = 5% p a, IRR = 12% with no reduction over time.

showed progressive depletion and that did not countenance a value higher than the original cost. This is perhaps understandable. It would be difficult to justify without good evidence. But it leaves distinct anomalies in the assessment to Capital Gains Tax on disposals of short leases where the amount of the statutory value of the depreciated leasehold interest can vary very considerably from its actual market value.[33] Reducing the period of depreciation would help to overcome this problem. As Sherlock Holmes remarked, 'It is a capital mistake to theorize before one has data.' The original legislators could not have hoped to have the necessary data available and their theoretical approach was substantially correct. Since the data are now available the theory should be amended to reflect the actual performance of short leaseholds in respect of their depreciation.

NOTES: CHAPTER SEVEN – CAPITAL GAINS TAX

1. *Turner* v. *Follett* (1973) S.T.C. 148.
2. Although the granting of a lease in return for rack rent would appear to be disposal in so far as the owner divests himself of rights in the asset, no liability occurs to CGT in practice.
3. Where land is compulsorily acquired on or after 6 April 1982 any gain realized by the acquisition can be deducted from the cost of other land bought to replace the original asset (FA 1982, s.83).

Capital Gains Tax has the dubious distinction of causing the identification of hysterectomy and castration as events causing disposal.

. Provided always that she was living with him for at least part of the year of assessment.

5. Circumstances might obtain where application of the new rules produces a gain while application of the old rules produces a loss. Or the reverse position could occur. In these circumstances the disposal will be considered to have resulted in neither a gain or a loss.

7. This may sound like common sense, but remembering that the interpretation of statute is necessary then the references are unclear (FA 1988 s.96(2)).

8. Clearly this will include any article, such as a painting, made by the tax payer. It also includes goodwill, copyright, patents etc.

9. *O'Brian* v. *Benson's Hosiery (Holdings) Ltd* [1977] 3 All E.R. 652.

10. Included in this exemption are vintage cars (which may have significant investment value).

11. For this purpose expenditure on the maintenance of a property outside the UK is considered to be personal expenditure.

12. A wasting asset is defined as having a life of fifty years or less (CGTA 1979 s.37). Note that the exemption for wasting assets does not extend to assets in respect of which capital allowances were given (or could have been given).

13. This exemption is clearly helpful to tax payers who have moved to another location and who experience difficulty in selling their house.

14. The proportion may be calculated by reference to the number of rooms or to the floor-space attributable to the different uses.

15. To obtain this additional exemption the let part can only be used for residential purposes.

16. Spouses living together could formerly claim an exemption in respect of a further private residence occupied rent-free by a dependent relative. However, this additional exemption has been withdrawn for disposals made after 6 April 1988 unless the dwelling was the sole residence of the dependent relative before that date (FA 1988 s.111).

17. A tax payer carrying on two or more trades is treated for these purposes as if s/he were carrying on one trade (CGTA 1979 s.115(7)). This will be so even if one trade discontinues and the assets are rolled-over to a different trade in the same ownership.

18. For these purposes land does not qualify if it is used or occupied for the purpose of dealing in land or the development of land (CGTA 1979 s.119). (But if a gain on the sale of the land is not in the nature of profit in the trading of land then it will qualify.)

19. This category of asset was identified for the first time in the Finance Act 1988 (s.112) and is in respect of acquisitions or disposals occurring after 27 July 1987.

20. Also included in FA 1988 s.112 and re. acquisitions or disposals after 29 October 1987.

21. A wasting asset is a depreciating asset, or one that will become a depreciating asset within ten years (CGTA 1979, s.117).

22. Although called retirement relief it is not necessary that the person actually go into retirement, only that the business assets are disposed of under the conditions required.

23. A person retiring for reasons of poor health is required to submit medical confirmation to the Inland Revenue (FA 1985 Sch.20 para.3).

24. *Harthan* v. *Mason* [1980] S.T.C. 94.
25. [1977] S.T.C. 206.
26. Excluded from the definition of such assets are shares held as investments.
27. A 'family company' is one in which either the individual has 25% of the voting rights or her/his near relatives have at least 50% of the voting rights.
28. The basis on which the chargeable gain is held-over is determined by the rules applying to gifts generally.
29. Freehold land and buildings are not wasting assets.
30. Leases with less than fifty years to run are treated uniquely as a special case and are treated separately for Capital Gains Tax purposes.
31. See, for example, *Clarke* v. *United Real (Moorgate) Ltd* [1988] S.T.C. 273 where the tenant met the landlord's costs of developing the site to be let in consideration of the grant of a lease. The amount was taxed as a premium.
32. See MacLeary, A. R. (1988) Irregularities in the taxation of short lease-holds, *Journal of Valuation* 6.4, pp. 351–364.
33. Although indexation and re-basing to 1982 should ameliorate the problem.

Chapter Eight

BETTERMENT TAXATION

THE CONCEPT OF RECOUPMENT

The recoupment of unearned increments in the value of land – known as betterment – is seen as the rise in value of a piece of land due to some external force that is not controlled by the owner of the land. But betterment of this kind can arise from different sources. It is convenient to attribute the accrual of betterment under two headings: general and specific.

General betterment arises as a consequence of macro-economic changes. In an expanding and intervention free economy land values will tend to increase in aggregate. Such rises in land value will give rise to capital gains so that recoupment of general betterment is liable to suffer taxation under Capital Gains Tax measures.

But when increasing land values arise in this way, that is by community and not individual effort, then there has always been a belief that such 'windfall gains' should be recouped entirely for the benefit of the community. Such a view was developed by neo-classical economists who believed that since demand, rather than any action of its owner determined the value of rent then this unearned increment should be taxable with the gain restored to the community.

These ideas were developed by writers such as Henry George and other 'single taxers'. They believed that the unearned increment in land value should be taxed at 100%. Such a tax was held, on its own, to be able to create sufficient revenues to permit the abolition of all other taxes.

Specific betterment arises as a result of Government intervention, particularly in land markets, and in an economic milieu therefore constrained in comparison with the intervention-free economies analysed by Ricardo *et al.*

Hence, in the United Kingdom, development cannot take place in an unfettered manner. Planning permission generally has to be sought (i.e., the permission of the state is required and such permission may be withheld). Again local planning authorities may decree land to be suitable for a higher and better use, thus creating a higher price for the land which has been so beneficially zoned (and always assuming a demand for the product of the development permitted). Hence, the state can directly increase the value of a particular parcel of land by a specific cause. Also the state can indirectly cause land values to rise by carrying out works of public expenditure on land which result in benefit to other land in private ownership. An obvious example of such land value increment accruing as a consequence of public expenditure occurs when access to land is improved by the construction of roads.

Specific betterment of this kind was recognized by the Uthwatt Committee when it identified it as, '. . . any increase in the value of the land . . . arising from central or local government action, whether positive, e.g., by the execution of public works or improvements, or negative, e.g., by the imposition of restrictions on other land.'

An illustration of the way in which specific betterment is realized and the amount which may legitimately be considered liable for betterment taxation may be useful. Let us assume a closed economic system within which the supply of housing land is fixed. Fig. 8.1 shows a level of demand (D_1) for this fixed supply (S-S) with a resulting value for housing land (V_1) at the intersection of the supply and demand curves. As the economy improves (to which the owners of housing land are not direct contributors) the demand for housing will increase as a consequence of general betterment. If we assume the demand rises to a new level (D_2) then the value of housing land will rise to V_2, equating with a

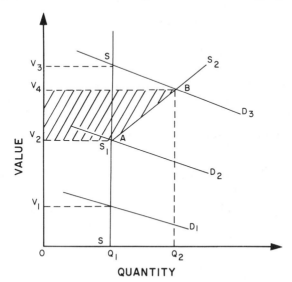

Figure 8.1:   The effect of land release and infrastructure provision and the scope for betterment recoupment.

point at the intersection of the (fixed) supply and (new) demand curves (A).

This trend of increasing land values will continue given demand fuelled by an expanding economy. If we assume a third level of demand (D_3) then we could observe that the price of housing land would rise to a figure represented by V_3 which equates with the interaction of the (fixed) supply curve and the (new) demand curve.

But suppose, before this third level of demand was reached, that state intervention took place to deal with the perceived social consequences of rising land prices. Such intervention would be the release of additional land for housing, no doubt serviced with infrastructure at community cost. This would introduce a new supply curve commencing at point A (S_1). The value of all the housing land on the market will now be increased to a figure V_4 which equates with the intersection of the (new) supply curve (S_1-S_2) with the demand surve (D_3) at point B and fixed by a new quantity of land Q_2.

Since the amount of building land is still controlled and therefore restricted, the amount of V_4 is greater than V_2 but less than the amount of V_3 which would have been realized without state intervention. The increment in value accrues to the owners of land who have received the benefit of planning permission as well as to the private owners of land who already have planning permission. Because of state intervention both the quantity and value of land have risen $(Q_1-Q_2$ and $V_2-V_4)$. Although it is the community alone which has created the enhanced value, and although the community may have had to invest to create the new development opportunities the community does not receive any return from its social action. On the other hand economic rent corresponding to the area of V_4, B, A, V_2 has been created, the sole beneficiary of which is the landowner(s).

Theoretically, therefore, this additional value may be taxed at 100% without affecting the supply of land. Equally no more than 100% of this additional value should be taxed since this would then further restrict the supply of land available.

LEGISLATIVE ATTEMPTS TO RECOUP BETTERMENT

These theoretical arguments for the recoupment of betterment have resulted in four attempts to create an equitable and workable system of taxation of windfall gains in land, all of which have been withdrawn.

The Town and Country Planning Act 1947 *et seq.*

The 1947 Act was the major piece of planning legislation in the United Kingdom. It included provision for the nationalization of development rights in land, following the recommendations of the Uthwatt Committee.[1] These provisions included a 100% charge which was levied on all specific betterment. By this means a person selling an interest in land would receive its existing use value while the state would recover any development value. The equity of this arrangement was maintained by arranging that compensation for land acquired compulsorily was also calculated on the basis of existing use value. Hence, the state was vested with all development rights (with owners of present rights to be compensated for the loss) while the only market for land was one which traded at existing use value. At one stroke the betterment recoupment problem had been solved. The solution was just and elegant, but simply because it was absolute and unavoidable it met resistance from those who might normally be expected to bring land forward for development. This attitude, encouraged by a belief that a change of Government would cause the scheme to be repealed led to a stagnant land market and restraint on economic development at a time when post-war reconstruction was urgently needed.

A newly elected government repealed the development charge in the Town and Country Planning Act 1954 thus removing entirely any recoupment of betterment. Unwisely compensation for land acquired by compulsion remained at existing use value. While this made it possible to bring forward land for public development more cheaply (and at a time of great social need) it created a dual-price market in land which was unfair and unstainable so that the Town and Country Planning Act 1959 had to include measures to restore market value to the code of compensation for compulsory purchase.

From a position of total recoupment of specific betterment from 1947 to 1954 there was a reversion to zero recoupment for either general or specific betterment from 1954 to 1965 when the Finance Act of that year introduced Capital Gains Tax. This act provided that tax be charged on all capital gains arising from the disposal of assets. Assets included all forms of property so that general betterment was taxed (at 30%) on all gains on interests in land (see Chapter 7, supra).

The Land Commission Act 1967

A betterment levy was introduced by this act which was designed to extract a proportion of development value when it was realized by transactions in land and from the development value that had been created by the actual development of land. The levy was intended to be a temporary measure designed to prevent a binary land market, as had occurred following the Town and Country Planning Act 1954, during the time that it would take for the Land Commission to buy all land needed for development. The rate of tax was 40% initially, but there was the intention of increasing the rate progressively to 45 and 50% at reasonably short intervals and with the possibility of further increases in the future.

The LCA was repealed in 1971. During its brief existence certain adjustments were found to be necessary in the provision of taxation measures for capital gains. These adjustments were intended to avoid any overlap between the tax paid under the FA 1965 and the LCA 1967. Essentially these charges were aimed at separating taxation of general betterment on current use value from taxation of specific betterment where land had development value.

The attempt at making this distinction was not altogether successful. It should be noted that a levy became due whenever a 'chargeable act or event' took place, whereas Capital Gains Tax remained payable only upon the disposal of an interest in the land.

Capital gains was again the only basis on which increments in land value were charged in tax between 1971 and December 1973.

The Finance Act 1974

With effect from 17 December 1973 'Development Gains Tax' was charged whenever there was a 'disposal' or 'notional disposal' of land or buildings with development value or development potential. The incidence of the tax, and the amount of chargeable gain, were derived from the application of a mathematical formula.

The taxable gain was the least of the following.

1. The disposal proceeds less 120% of the cost.
2. The disposal proceeds less 110% of the current use value at the date of disposal.
3. The full gain less the increase in current use value over the period of ownership, or since 6 April 1965, where land was owned before that date.

In making these calculations a 'threshold' of £10 000 (£1000 in the case of companies) with relief up to £20 000 (£2000 in the case of companies) was allowed. Gains calculated in accordance with this formula were taxed at Corporation Tax rates in the case of companies, and at Income Tax rates in the case of individuals. Any gains not subject to Development Gains Tax under the formula would be subject to Capital Gains Tax. In effect general and specific betterment were being distinguished with the charge to tax being made first on the specific and then on any general betterment remaining.

DGT became chargeable where there was a disposal of the tax payer's interest in the land and buildings concerned. In addition, a chargeable event occurred where material development had been carried out, and the buildings were subsequently let. In these circumstances the 'first letting' was to be treated as a disposal for the purpose of taxation and as giving rise both to Capital Gains Tax and to DGT.

DGT was superceded from 1 August 1974 by Development Land Tax (DLT).

DEVELOPMENT LAND TAX

Attempts to tax betterment are interesting not only in themselves, where theory and practical means can sit uncomfortably with each other, but also because a betterment tax is the only item in the national taxation armoury which addresses itself exclusively to land.

Efforts to weld theory to practice, and the fact that the tax requires an understanding of notions and actions not required in other areas of taxation, make it compulsively interesting to those who busy themselves with land management. Among professional groups, Chartered Surveyors are specially competent to take an inclusive view of the various facets of economic and legal understanding and of valuation and town planning practice. Since the notion of betterment is correlative to worsenment, and since surveyors are again familiar with the law and practice of valuation for the compulsory acquisition of land, then they are well equipped to deal with the concepts upon which Development Land Tax has been based and the necessary assumptions and interpretations which have to be made to prepare the various valuations required in order to reach the statutory quantification of chargeable realized development value.

Before proceeding to see how the Development Land Tax attempted to quantify realized development value it will be useful to consider briefly the more significant concepts which the DLT Act required to be brought into play.

a) Interests in land

An 'interest in land' for the purposes of DLT has its normal meaning including any estate or interest in land and any right in or over land. Hence 'interest in land' embraces not only major interests in land such as freeholds and leaseholds but also rights affecting the use of land or the disposition of land. Therefore not only are restrictive covenants, as lesser interests, included but so also are rights accruing under a contract to acquire interests in land.

Interests in land are central to DLT since the tax is based on the disposal of such interests. (This may be compared with CGT where the tax is based on the disposal of an 'asset'. An asset with respect to land will be an interest in such land.)

Special provision in DLT legislation apply wherever an interest in land has been 'assembled' by the acquisition of different interests in land, possibly at different times. In these circumstances the cost of acquisition of the interest being disposed of would be taken as the aggregate cost of all the constituent parts of that interest on the assumption that each interest was acquired upon the date of the last interest purchased.

Special rules also applied to the grant of an option to purchase an interest in land. On the face of it this is a disposal of part of the owner's rights in land. DLT required an option to be regarded as the creation of a new interest in land.

Interests in land were therefore sometimes modified to suit the specific betterment legislation, although in the main an interest in land and buildings remained exactly what they were in law. A person who developed a building on a piece of land in which he had the freehold interest was merely carrying out improvements on that land. No new interest was, or needed to be, created. However, DLT applied not only to the actual disposal of interests in land, but also to the realization of development value brought about by the commencement of a project of material development. For the purpose of DLT the fiction was devised whereby the owner of the 'major interest' in the land was held to have sold and then immediately re-acquired his interest at the same price. The price was deemed to be full market value. The owner of such a major interest was also deemed to hold all interest in the land subject only to some minor exceptions.

b) Material development

When a project of material development was commenced on land the DLTA required that every major interest in the land was deemed to be disposed of and immediately reacquired as has just been explained. The identification of material development was clearly important for this reason. It was also important for the further reason that the current use value of the major interest in land had to be determined with reference to market value but on the assumption that it would be unlawful to carry out any further material development. Since the quantity represented by the current use value is one of the items used in determining the amount of the realized development value then the identification of material development was again important.

The statutory definition of material development was taken from existing planning legislation[2] (DLTA 1976, ss.7(7), 47). Hence, material development is any development other than that for which planning permission was granted by a general development order (GDO) currently in force. If any building or engineering operation or any change of use was within the definition of 'development' as defined in the planning acts then it could not be material development for the purposes of DLT. Due regard would require to be had of the GDO (and any conditions or limitations expressed in the GDO) or to conditions attached to any planning permissions.

However, the DLT Act specifically excluded certain building works and changes of use which were not to be regarded as material development (DLTA 1976 Sch.4 Part II). It is not appropriate to list here those excluded operations or changes of use other than to say that they were generous in so far as the cubic content of a building could be increased by up to one third before it would be regarded as material development (after March 1981). Again where a development extends to two or more buildings within the same curtilage they could be regarded as a single building (DLTA 1976 Sch.4 para.5(3)(a)). Much of the value in these specified exclusions is that they served to give a clearer idea as to the kind of operations that were considered to be material development.

Subject to these exclusions the concept of development and, more particularly, 'material development' can be seen as one which was entirely and purposefully rooted in town planning legislation. In the United Kingdom town planning measures are the means of expressing policies based on welfare economics. By this means development rights are preserved to the state and released by government intervention through the granting of planning permission. Where development value accrues it can therefore be clearly identified (if not quantified) with references to building operations or changes of use on specific parcels of land.

c) Disposal

Under the DLT provisions a tax payer was only liable for assessment and payment of tax when a disposal of an interest in land took place. The Act did not define actual disposals which would appear to have been self-evident when a transfer of an interest in land took place between one person and another. (However, certain disposals which were actual disposals such as gifts of interests in land were not treated as disposal for the purposes of DLT. Since there were no proceeds, no development value was realized and therefore no liability to DLT arose.)[3]

However, the concept of disposal was extended to include deemed disposals and part disposals.

A deemed disposal occurred at a point in time immediately before the commencement of a project of material development on land, (where an interest in land was not otherwise actually conveyed.)

A part disposal may be a physical separation of part of the land, or it may refer to a disposal of part of the total (or major) interest in the land, such as the granting of a lease, whereas such different kinds of disposal are distinguished from each other for the purposes of CGT, they are treated the same for the purposes of DLT. Part disposals can therefore arise in a variety of ways and forms. For example, if an owner of an interest in land received a compensation or insurance payment he would have been treated as having made a part disposal of his interest (DLTA 1976 ss.3(1) and (2)). In such cases, therefore, there could have been a deemed part disposal.

For most purposes the date of disposal for DLT purposes was the date upon which an unconditional contract was entered into. In the case of a deemed disposal its date was assumed to be at a moment in time immediately before the commencement of a project of material development. Any major interest in land was deemed to have been disposed of and re-acquired at that moment (DLTA 1976 ss.2(1) and 45(3)). There was irrefutable logic in the selection of a date closer to the point when development would be created (the grant of a planning permission) than the completion of development, when such unearned income may have been realized. On actual disposal the owner of any major interest would have benefited from the gain at the point of sale. Where gains are not actually realized, however, it is more difficult for the tax payer to meet any liabilities due to the Inland Revenue. However, it should be noted that a tax payer undertaking a project of material development on land which he owns will have an asset of enhanced value against which he will be able to borrow. Additional difficulty caused by this provision was that an assessment of development value made at the commencement of a project must be very much less certain than one made on completion when the value of the product can be measured with the knowledge of actual conditions obtaining.

d) Current use value
We will find, when we look at the mechanism for calculating DLT that current use value forms a significant part of the calculation designed to quantify realized development value.

The act defines current use value of an interest in land as the amount which the interest would realize if offered on the open market but on the assumption that it will remain in its present use in perpetuity and that there is no permission to carry out any material development of the project (DLTA 1979 s.7(2)).

The same section of the act allows for assumptions to be made that planning permission would be granted for 'existing use development' (DLTA 1979 s.7(2), Sch.4). This provision again mirrors those activities specified in the planning acts as coming within existing use and not requiring planning permission.

THE TAXATION SCHEME

Development Land Tax (DLT) was based on the realized development value accruing to the tax payer in any fiscal year. The net chargeable realized development value was the amount upon which tax was payable. For this purpose all development value which was realized was chargeable unless there was statutory provision for a specific exemption from the tax.

The DLT which was paid was the surplus which the tax payer retained after defraying any costs incidental to the realization of development value, such as legal costs, valuation fees and Stamp Duty, and deducting a statutorily defined 'relevant base value' (see infra).

Consequently the tax payer was not paying tax on a development value established by a valuation. Rather the tax payer was paying tax on a development value which the legislation defined by reference to a surplus depleted by allowable deductions. Such a sum need not necessarily bear any resemblance to the amount of development value ascertained by a valuer. As with most taxation legislation, the means by which development value is calculated is exclusively that which the legislation allows, and which is based on a clear enough concept of development value, not on any alternative approach even if such an approach should produce a more equitable result.

These comments should perhaps be leavened by the observation that in the particular case of DLT every attempt had been made to learn from legislative mistakes made in the past. Where there is an element of doubt or judgement it is probably fair to say that this is usually (although by no means exclusively) construed in the tax payer's favour.

The DLT Act was introduced co-terminously with the Community Land Act of 1976. Together they made up the Community Land Scheme the justification of which were set out in the White Paper *Land*.[4] The scheme had two objectives.

1. To enable the community to control the development of land in accordance with its needs and priorities.
2. To restore to the community the increase in value of land arising from its efforts.

The scheme was therefore an all-embracing approach to development including provisions for 'positive planning' by local authorities, or as might be said today a proactive rather than a reactive role, whereby local authorities would be able to identify and implement development opportunities. The means for securing this first objective were contained in the Community Land Act.

The provisions for the recoupment of betterment where contained in the DLTA and were originally intended to be temporary. At some future date (the 'second appointed date' in the CLA) all development rights would be vested in the state and land could only change hands at existing use value.[5] During the transitional period, however, it would be necessary to have measures to enable the community to reap the benefits of development value.

By preparing companion legislation for the recoupment of betterment the legislators of the day were able to create taxation measures which were capable of standing on their own should the CLA founder as a result of the political controversy surrounding it. Indeed this is what actually happened. The CLA was disposed of as soon as Conservative government took power in 1979, but the DLTA survived until 1985.

THE BASIS OF CALCULATION

The calculation for DLT purposes can be expressed as follows.

	Proceeds of disposal	=
less	costs of disposal	= _____
	Net proceeds of disposal	=

Deduct the highest of

Base A – the aggregate of
1. Cost of acquisition of the interest in land.
2. Cost of 'relevant' improvements.*
3. Increase in CUV since purchased (or 6 April 1965).†
4. Special addition to reflect interest charges.‡

Base B – the aggregate of
1. 115% of CUV at date of disposal.§
2. 100% of expenditure on 'relevant' improvements.

Base C – the aggregate of
1. 115% of cost of acquisition of the interest.§,‖
2. 115% of all improvements.§,‖

| | *less* | Highest base value | = _____ |
| | | Realized development value | = _____ |

---

*Relevant improvements consists of actual expenditure less an amount by which CUV at disposal has been increased as a result of these improvements.

† CUV = Current Use Value: the April 1965 date applied if it was the later of the two dates.

‡ If the interest in land was acquired before 1 May 1977 a particular relief was allowed for the high interest rates which would have obtained on the cost of acquisition but up to a maximum of four years.

§ 110% for disposals before 25 March 1980.

‖ 150% for a deemed disposal after 9 March 1981 on the start of development for private dwelling – where land was held as stock.

Two matters become obvious when considering the basis of calculation for DLT. Firstly, it can be seen that only one Base (B) appears to attempt to address the question of what the actual development value might be. With minor adjustments it will produce a net figure which is approximately the difference between the market value on disposal and the current use value at the same date, thus attempting to quantify specific betterment. What then is the function of the other two Bases? Essentially they appear to have been designed to ensure that if the history of any case reveals no profit, then there can be no liability to DLT. It is not immediately obvious which Base might have proved the more favourable, and each Base value would need to be calculated on each occasion that an attempt was made to quantify realized development value.

This brings us to the second matter which will occur to a valuer when considering the basis of calculation, and that is the number of different valuations which will require to be carried out. As has been mentioned the DLTA does not require a valuation of development value, it requires the calculation of a variety of allowable deductions which may be made from the proceeds of disposal. If the disposal is itself deemed then a value will have to be generated in the absence of an actual amount. In any given case therefore a surveyor may have to generate the following valuations.

1. The value of the major interest in land on deemed disposal.
2. The current use value on the date of disposal.
3. The current use value on the date of acquisition.
4. The increase on current use value at the date of acquisition as a result of improvements upon date of disposal.

Example 8.1. The basic DLT calculation for total disposal

A sold land and buildings in 1984 for a sum of £1 500 000. Its current use value at the date of sale was £1 000 000. The incidental costs of disposal amounted to £20 000. The property was purchased in 1978 for a sum of £500 000 which was its current use value at that time. Between purchase and sale £250 000 was spent on various improvements of which £150 000 was spent on qualifying relevant improvements.

The calculation of realized development value would proceed as follows.

		£
Proceeds of disposal	=	1 500 000
less costs of disposal	=	20 000
Net proceeds of disposal	=	1 480 000

BASE A*
1. Cost of acquisition   500 000
2. Expenditure on relevant improvements   150 000

3. Increase in current use value

CUV at disposal	1 000 000	
less CUV at acquisition	500 000	500 000
		1 150 000

BASE B

1. 115% of CUV at disposal

500 000 × 1.15		575 000
2. Expenditure on relevant improvements		150 000
		725 000

BASE C

1. 115% of cost of acquisition

1 000 000 × 1.15		1 150 000
2. 115% of expenditure on improvements		
250 000 × 1.15		287 500
		1 437 500

less	highest Base Value (Base C)	=	1 437 500
	Realized development value	=	42 500

---

* Because the interest was purchased after 30 April 1977 the Special Addition to Base A is not available.

PART DISPOSAL

We have seen that a part disposal arose where the owner of property granted a lease or any other interest out of her/his own interest or where she/he disposed of some but not all of her/his interest in land to another person. We have also seen that part disposal can occur in situations where a compensation payment is paid, an insurance claim is paid or any other capital sum is received. In these circumstances the owner of the major interest in land was treated as having made a part disposal of the property to which such benefits related.

The entire interest in land immediately prior to part disposal was described in the legislation as 'the relevant interest'. (DLTA 1976 Sch.2 para.9). The interest disposed of by the part disposal was known as 'the granted interest' (DLTA 1976 ss.3(5)(b) and 47(1)) and the interest remaining after the disposal as 'the retained interest' (DLTA 1976 ss.3(5)(a) and 47(1)).

For the purpose of part disposal the rules for total disposal broadly applied. The net proceeds of disposal were calculated in exactly the same way. The acquisition factors were apportioned by reference to a formula (infra) while the current use value was ascertained by special rules (infra). Further special rules applied where an actual disposal followed shortly after a deemed disposal and where the part disposal took place at less than market value.

The method used for apportionment was to take the full cost of acquisition and the full amount of expenditure on improvements and relevant improvements relating to the relevant interest in land and then to attribute to the part disposal an appropriate fraction of such costs (DLTA 1976 s.3 and Sch.2 para.10(1)).

The gross figures were reduced by applying to them the fraction

$$\frac{PD}{PD + MR}$$

where PD was the net proceeds received from the part disposal; and MR was the market value of the retained interest immediately after the disposal.

Should it have been possible to determine the current use value (CUV) at the time of disposal, then that amount would have been used in the DLT calculation (DLTA 1976 s.3 and Sch.2 para.12(4). In the absence of such information the CUV was to be determined by reference to the formula

$$CW - CR$$

where CW was the current use value for the whole of the property immediately prior to the part disposal; and CR was the current use value of the retained interest after the disposal (DLTA 1976 s.3 and Sch.2 para. 12(1)).

In this manner CUV at acquisition was to be determined by reference to the formula, (DLTA 1976 s.3 and Sch.2 para.12(2))

$$\frac{CW - CR}{CW}$$

The CUV at the time of disposal is then determined by applying the above formula to the CUV at acquisition to derive a CUV for the part disposed of at the time of acquisition.

CUV of part disposed at acquisition = CUV (of relevant interest at acquisition) × (CW – CR/CW).

Example 8.2. Calculation for physical part disposal

A acquired a freehold property in 1980 at a cost of £300 000 (when its CUV was £49 000). He spent £5 000 on obtaining planning permission to develop part of the property. In 1985 A sold that part of the property with the benefit of planning permission for a sum of £200 000 after incidental costs of sale. At the date of sale the market value of the retained interest was £160 000. The CUV of the relevant interest was £70 000 and the CUV of the retained interest was £30 000. A calculation of the chargeable realized development value is required.

Before proceeding to the calculation it will be necessary to derive CUVs for the part disposal at both acquisition and disposal where such amounts are otherwise unknown. Hence

CUV attributable to the part
    disposed, at disposal

$$= CW - CR$$
$$= £70\,000 - 30\,000$$
$$= £40\,000$$

CUV attributable to the part
    disposed, at acquisition

$$= CUV \times \frac{CW - CR}{CW}$$

$$£49\,000 \times \frac{70\,000 - 30\,000}{70\,000}$$

$$= £28\,000$$

The calculation of chargeable realized development value would then proceed as follows.

Net proceeds on part disposal $= £200\,000$

*Base A**

1. Cost of acquisition of part $= \text{cost} \times \dfrac{PD}{PD + MR}$

$$= 300\,000 \times \frac{200\,000}{200\,000 + 160\,000} \qquad = 166\,667$$

2. Expenditure on relevant improvements $=$ 5 000†
3. Increase in CUV of part

CUV at disposal	40 000‡	
less CUV at acquisition	28 000‡ $=$	12 000
	$=$	183 667

Base B
1. 115% of CUV of part at disposal
    40 000‡ × 1.15 $=$ 46 000
2. Expenditure on relevant improvements $=$ 5 000
                              $=$ 51 000

BASE C
1. 115% cost of acquisition of part
    (From Base A)   166 667 × 1.15 $= 191\,667$

2. 115% expenditure on improvements

5 000 × 1.15	=	5 750
	=	197 417
less highest base value (Base C)	=	197 417
Realized development value	=	2 583

---

* Because the interest was purchased after 30 April 1977 the Special addition to base A is not available.

† The cost of obtaining planning permission is a relevant improvement, it is not apportioned because the expenditure was entirely in respect of the part disposed.

‡ Previously calculated.

If A sold the remaining interest in the property for £160 000 (net) the chargeable realized development value would be calculated as follows.

			£
Net Proceeds on disposal			160 000

Base A

1. Cost of acquisition	300 000		
less: allowed on part			
disposal	166 667 =	133 333	
2. Cost of relevant improvements since part disposal		0	
add: prior			
improvements	5 000		
less: allowed on part disposal	5 000 =	0 =	0
3. Increase in CUV			
CUV at disposal		30 000	
less: CUV at acquisition	49 000		
– allowed on part disposal	28 000 =	21 000 =	9 000
			142 333

Base B

1. 115% of CUV at disposal	30 000 × 1.15 =	34 500	
2. Expenditure on relevant improvements	=	0	
		34 500	

Base C

1. 115% of cost of
   acquisition (from
   Base A) $133\,333 \times 1.15 =$ 153 333
2. 115% of expenditure
   on improvements $=$ 0
                                                        153 333

  less   highest base value (Base C) $=$ 153 333

      Realized development value $=$ 6 667

LEASES AND REVERSIONS

The grant of a lease was a part disposal for DLT purposes. The liability to DLT on the grant of a lease was therefore dependent upon the part disposal rules. But there were also special rules for leases and these required an understanding of further concepts which are discussed below.

A part disposal in respect of a lease could be effected by means of a capital sum (premium). But, more commonly the lease would be arranged through the receipt of rent over a period of years specified in the agreement. Should development value be realized through the receipt of rent, then, theoretically that part of the rent reflecting development value should be subject to betterment taxation. Under the DLTA, however, no part of the rent was subject to taxation. An alternative arrangement was adopted whereby development value was conceived of as having been entirely realized at the date of the granting of the lease. This fiction was implemented by introducing two artificial events. Firstly the landlord was assumed to have received an amount equal to the capitalized value of the future stream of rental payments due to be made throughout the duration of the lease.[6] The landlord could then be made liable to the imposition of a charge to tax under the DLTA in respect of any development value identified in that capital sum. Secondly, the tenant had to be treated as having paid the capitalized value of the right to an income under the terms of the lease.

A consequence of introducing this artificiality in the legislation was that it was then necessary to make corresponding adjustments, in respect of fictional payments, each time there was a dealing with the lease, or with the reversion.

The position is further complicated by the requirement that 'the landlord's rental rights' under a lease, from which the capital sum required by the legislation was derived, refers inclusively to the right to receive rent during the term of the lease but excludes the right to recover possession of the land on the expiry of the lease (DLTA Sch.2 para.13(2)). It is the value of the term plus the reversion which comprise the monetary worth of the landlord's interest in normal circumstances, but this separation is again necessary in order that the logic of the original fiction may be pursued.

The application of the capitalization of the rental payments and of separation of the value of the reversion from the landlord's rental rights can be demonstrated by reference to an illustration.

Assume that A owns the freehold of a property which he leases to B. The value of A's remaining interest (i.e., the term plus the reversion) is £600 000 of which £100 000 is the value of the reversion and £500 000 is the value of the 'landlord's rental rights' (i.e., the right to receive income during the term of the lease and as defined by the DLTA). On the supposition that A is liable to DLT in respect of development value 'realized' by this part disposal then it will be calculated on the basis of the 'receipt' of the £500 000 being the capital value of 'the landlord's rental rights'. The value of the reversion would not be assessed for DLT.

Now assume that A sells his rental interest in the land to C for a sum of £750 000. In order for the realized development value to be identified for that disposition the DLTA will require that the £500 000 attributable to the landlord's rental rights will be deducted from the proceeds of sale so that A will be treated as having received £250 000. A is therefore accountable for DLT for the total consideration which he receives, but at different times.

The logic of the scheme requires that C is treated for DLT purposes as having acquired A's entire interest in land for £250 000 (i.e., the same consideration that A received for the purpose of DLT). The same logic requires that B paid A the sum of £500 000 at the time that the lease was granted. Should B decide to assign his lease to D, therefore, he will be treated as having received a sum of £500 000 when the lease was granted.

Example 8.3 Calculation for part disposal on the grant of the lease

A acquired the freehold interest in town centre property in 1980 for the sum of £150 000 when its current use value was £125 000. In 1984 he let the property to B at a rent of £30 000 per annum for a term of twenty-one years. The rent reflected the fact that B was able under the lease to convert the property to a higher and better use, and consequently the rent was higher than would have been expected for the existing use. Immediately before the lease was granted the property had a market value of £300 000, and a current use value of £200 000. The retained interest has a market value of £250 000 and a current use value of

£230 000. The landlord's rental rights are calculated as being worth £225 000.

The calculation of chargeable realized development value would then proceed as follows.

Base A

1. Cost of acquisition of part = cost × $\dfrac{PD}{PD + MR}$

   $= 150\,000 \times \dfrac{225\,000}{125\,000 + (250\,000 - 225\,000)}$ $=$ 135 000

2. Expenditure on relevant improvements $=$ 0

3. Increase in CUV of part
   CUV at disposal = CW − CR = 200 000 − (230 000 − 225 000)

   $= 195\,000$

   less CUV at acquisition = CUV × $\dfrac{CW - CR}{CW}$

   $128\,000 \times \dfrac{195\,000}{200\,000}$ $=$ 124 800 $=$ 70 200

   205 204

Base B

1. 115% of CUV of part at disposal = 195 000 × 1.15 $=$ 224 250
2. Expenditure on relevant improvements $=$ 0

   224 250

Base C

1. 115% of cost of acquisition of part = 135 000 × 1.15   155 250
2. 115% of expenditure on improvements $=$ 0

   155 210

   Proceeds of part disposal (i.e., the landlord's rental rights) $=$ 225 000

   less highest base value (Base B) $=$ 224 250

   Realized development value $=$ 750

CONCLUSIONS

The reader who has made it this far may feel that he has little appetite for further enquiry into the nature of DLT, as the latest example of betterment taxation. Happily, it is not appropriate in this text to further detail the provisions and workings of development land tax legislation and computations. Sufficient to say that the workings of the DLTA are much more comprehensive than indicated here and often very complicated in application.

The intention of this text is not to spell out the various manifestations of DLT as a means of recoupment of development value to the community but rather to use DLT as an illustration of the problems inherent in attempting to make equitable provision for such taxation.

As we can see, it is not the concept of betterment which gives difficulty, it is its application. Such difficulty can be categorized into three areas.

Firstly, it is clear from the history of betterment recoupment measures that they are prone to change in a democratic society where the complexion of government can change frequently, and the view of societal expectation and the management of the economy with it. Any truly successful measure of recoupment would need some measure of confidence that it would be sustained at a comprehensive level before it could hope to become workable and robust. Given political uncertainty, those tax payers likely to be affected by such measures are going to have particular regard to the perceived durability of any measure and plan their affairs accordingly.

Secondly, the scope and purpose of the recoupment method itself needs to be acceptable. The history of attempts to tax betterment would suggest that a real world attempt to confiscate 100% of unearned increment is unlikely to meet the necessary test of political acceptability. Equally, the means of distributing the tax take have to be politically acceptable at both national and local level. The Betterment Levy experience showed that the will to link planning measures with betterment recoupment is weakened if the fiscal benefit is only at the national level and not to some degree at the local level where most initiatives or community actions leading to specific betterment take place.

Thirdly, whatever measures are introduced they should be capable of being understood in respect of their detailed application and also capable of being implemented in the sense that the figures which are required to be generated to calculate development value should be ones which could reasonably be expected to be generated in the real world. The requirements for valuations within recoupment measures can be such as to regress the calculation back to hypothesis instead of forward into acceptable practicality.

It is clear that each succeeding legislative attempt to recoup betterment in the United Kingdom has learnt something from the mistakes of its predecessors. The DLTA even survived the demise of the Community Land Scheme, its architects having learnt that it was a useful contribution to the robustness of the taxation aspect if it were capable of being decoupled from any specifically detailed planning measures, such as in this case, 'positive planning'.

Although the DLTA was a survivor of six years of Conservative government, and although an indication of its then general acceptability was the universal exclamations of surprise when the Chancellor of the Exchequer announced its demise, it actually foundered for 'external' reasons rather than for any real concern about its internal workings. Partly these external reasons had to do with reforms of fiscal policy, but also with the sober realities of the effectiveness of DLT measured against its revenue (both absolutely and relatively) and to its cost of collection. A complicated tax requires a bureaucracy and carries many additional hidden costs in its collection. On the other hand DLT's contribution to the national coffers could only be measured in tenths of one per cent.

The justification for this chapter is, nevertheless, that the taxation of betterment is wholly equitable. It is one of the few taxes that takes from unearned income rather than the earned income. Because it is exclusive to land it is ultimately related to planning and land use measures which may regain significance in an increasingly environmentally conscious world where Green is expanding its share of the political rainbow. Students and practitioners therefore have to be aware of the theory and some of the problems relative to betterment recoupment in order to prepare themselves for its future re-emergence, in whatever guise.[7]

NOTES: CHAPTER EIGHT – BETTERMENT TAXATION

1. *Report of the Expert Committee on Composition of Betterment*, Cmnd. 6386, HMSO, 1942.
2. Town and Country Planning Act 1971, Town and Country Planning (Scotland) Act 1972 and Planning (Northern Ireland) Order 1972.
3. Note, however, that the donor and the donee may have been liable to other taxes such as Inheritance Tax.
4. *Land* Cmnd. 5730, HMSO, London.
5. c.f. the financial provisions of the TCPA 1947 – but with a transitional stage to avoid the difficulties experienced with the 1947 Act.
6. If the lease was subject to a premium together with rental payments then the two capital amounts would be summed.
7. A significant expression of the re-awakening of interest in the analysis of public and private costs and benefits relative to land use and the environment can be found in *Blueprint for a Green Economy* (1989), by D. Pearce *et al.*, published by Earthscan Publications.

Chapter Nine

LAND VALUATION

INTRODUCTION

In pursuing the objectives of the taxation of land it will be necessary to value an interest in land where a taxable event occurs which has a bearing on that interest. The land may be acquired or disposed of. The consideration may be for cash, or it may be by way of a gift, or a sale under value partly for cash and partly by way of gift. It may be by way of exchange of land. Land may be deemed to be disposed or a hypothetical sale may be assumed. Given the nature of land and the complex web of interests which can obtain in property, then allied to the remorseless logic of taxation law and practice, a very large number of different taxable situations can arise. Often such a taxable situation can be in circumstances where it is not always clear that the actual consideration represents market value of the asset for the particular tax.

We have already seen that with taxes on income and expenditure special cases arise where the valuation of land and buildings have to be treated in a special or prescribed manner. In the case of the capital taxes the significance of special taxable situations for the value of property assets is even greater because of their very nature. Each tax carries its own particular requirements regarding the valuation of assets, but these requirements are usually only variations on a general concept. Before proceeding therefore it will be useful to consider what the general position regarding the valuation of land and buildings for taxation (and other purposes), is, so that these differences between taxes can be seen in context.

VALUATION AND THE LAW

Where valuation occurs as a requirement of legislation then clearly there is required to be a legal value for that purpose, whether the purpose be taxation, compulsory acquisition, arbitration or any other activity sanctioned by statute. In the absence of the requirement to establish a legal value, the valuation of land and buildings is not a 'legal' activity, in the sense that the law is disinterested in any bargain struck between parties to a transfer of an interest in land or any valuation of that interest which may be made for the purpose of the conveyance, provided always that a valuation is not required for a particular statutory purpose.

Unlike in most countries, those who may carry out such valuations in the United Kingdom are not required to be registered with the state, although professional accountability can be significant.

Again, as all valuers are aware, a figure derived as a valuation is invariably based on assumptions some of which may include an assessment of future events. Therefore any valuation is, by its very nature, uncertain. Market value of an interest in property is only therefore established when an actual transaction occurs and where the conditions are such that no undue peculiarities in the circumstances have disturbed the consideration paid, which can then be taken as representative of value.

The state has long recognized the reality and the potential complications of attempting to establish the value of an interest in land in the absence of an 'arm's length' transaction. The history of legislative provision and interpretive case law and the attempt to establish equity with respect to the value of land for compulsory acquisition pay testimony to the difficulties inherent in the process.

In the case of the valuation of land for the purpose of taxation then, outside of particular legislative instruments, the basis for the determination of value has been broader. This general basis provides the basic rules which run through all those measures effecting the taxation of land.

MARKET VALUE

The basic rule is that the valuation of land for taxation purposes shall be 'market value'. Market value of an interest in land is held to be the consideration which an interest might reasonably expect to fetch on sale at the relevant time in the open market. A definition of this kind is found in Section 7(1) of the Development Land Tax Act 1976, Section 150, and Sch.6 of the Capital Gains Tax Act 1979 and Sec.38(1) of the Finance Act 1975, (in respect of Capital Transfer Tax and Inheritance Tax).

Subject to the particular rules for these various taxes there is a general understanding therefore that the valuation of an interest in land for taxation purposes shall be at 'market value'. The original

definition of the value of land for taxation purposes lies in the Finance
Act 1894, Sec.7(5). This dealt with the 'principal value' of the property
passing on death for estate duty purposes and defined this value as the
price which, in the opinion of the Commissioners of Inland Revenue, the
property would fetch if sold in the open market at the time of the death of
the deceased.

Subsequent case law and legislation has reaffirmed the intention of
this legislative aim.

INTERPRETATION OF MARKET VALUE

A leading judicial comment on the concept of market value in respect of
any interest in land was provided by Swinfon-Eady L. J. in IRC v. Clay
and Buchanan.[1] It was his opinion that the legislation required, in
respect of the value of an interest in land, its ascertainment by reference
to the amount obtainable in an 'open market' which included every
possible purchaser.

> The market is to be an open market as distinguished from
> an offer to a limited class only such as the members of the
> family, the market is not necessarily an auction sale. The
> section means such amount as the land might be expected to
> realize if offered under conditions enabling every person
> desirous of purchasing to come in and make an offer and if
> proper steps were taken to advertise the property and let all
> likely purchasers know that the land is in the market for
> sale. It scarcely needed evidence to inform us – it is common
> knowledge – that when the fact becomes known that one
> probable buyer desires to obtain any property that raises the
> general price or value of the thing on the market. Not only is
> the probable buyer a competitor in the market but other
> persons such as property brokers compete in the market for
> what they know another person wants with a view to resale
> to him at an enhanced price so as to realize a profit. A
> vendor desiring to realize any land would ordinarily give
> full publicity to all facts within his knowledge in order that
> he might thereby obtain the amount that the land might be
> 'expected to realise'. All these matters ought to be taken into
> consideration. 'Expected' refers to the expectation of
> properly qualified persons who have taken pains to inform
> themselves of all the particulars ascertainable about the
> property and about its capabilities, the demand for it and
> the likely buyers. The price actually realized by a sale is not
> necessarily the price which it might have been expected to

realize but if the value be competent there ought to be little
difference between the two figures.

Certain further detailed aspects of market value have to be taken
together with this broad concept.

If it is proved that the value of property has been depreciated by
reason of the death of the deceased then the Commissioners, in fixing
value, are charged with taking such depreciation into account.
(F(1909–10)A 1910 s.60(2)). This might be the case, for example, where
the goodwill of a business was significantly diminished because the
deceased was a proprietor.

Where money has been expended after the death with a view to
enhancing the price obtainable, some allowance may be made for such
expenditure.[1] This judgement is consistent with the proposition that a
sale should be arranged in such circumstances that the property is made
as attractive as possible to prospective purchasers.

In Ellesmere (Earl) v. IRC it was held that the 'open market' value
assumes a sale made 'in such a manner and under such conditions as
might reasonably be calculated to obtain for the vendor the best price for
the property'.[3]

If a sale in lots would produce a better price, sale of the property as a
single unit is not to be presumed, and this principle applies even
although sales in lots would take several years to complete.[4]

In Duke of Buccleugh v. IRC it was also confirmed that the expenses of
selling are not deducted, (unless the property is a sale of an estate out of
which the administration has not yet been completed), since in the
market place no adjustment would normally be made for the vendor's
expenses.[5]

It is of no consequence that the property passing could not in fact be
sold. One is still required to arrive at the value which a willing buyer
would pay a willing seller. Thus in estimating market value no deduction may be made for the fact that the whole of the property is assumed
to be placed on the market at one time. (F(1909–10)A 1910). This
provision is commonly referred to as the 'flooded market clause'.

As we have seen, within the general concept of market value as
considered by Swinfon-Eady a special purchaser is not excluded. The
idea that the open market and any purchaser, including a special
purchaser, falls to be included were considered prior to his judgement in
IRC v. Marrs Trustees[6] and subsequently in Glass v. IRC.[7]

However, the concept of a hypothetical purchaser including a special
purchaser is one that has continued to give difficulty as can be seen in Re
Ashcroft Clifton v. Strauss and in Re Lynall v. IRC.[8,9] The concept of the
hypothetical vendor has been somewhat less contentious but has still
provided its own particular difficulties in interpretation.[10,11,12]

Finally case law has confirmed, that for taxable transfers brought about my death, the date of the valuation is the date of death.[13,14]

PROFESSIONAL PRACTICE AND OPEN MARKET VALUE

When considering the basic concept of open market value for the purposes of land taxation, and remembering that the genesis of legislative definition and interpretive case law is mainly historic when land ownership was restricted to a few privileged hands and when, more importantly, land markets were very much less restricted by public regulation, then it may be useful to examine at the same time the concept of open market value as seen and practised in the everyday world by valuers, (and for that matter by accountants and those dealing with securities.)[15]

Over the last decade there has been a move in the surveying (and accountancy) profession to introduce a homogeneous approach to the valuation of property assets, particularly where historic costs may bear little relationship to current market value, especially in times of high inflation. Clearly such a situation has implications for accountancy practice but it has certainly focused the mind of landed professions, such as the Royal Institution of Chartered Surveyors, on the need for agreed and practical guidelines in respect of asset valuation standards. These have been produced and are in harmony with concerted international adoption of these standards. Included in these standards is the definition of 'open market value'.[16] The complete definition is reproduced below.

Land and Buildings

Definition of Open Market Value and Forced Sale Value

1.1 'Open Market Value' means the best price at which an interest in property might reasonably be expected to be sold at the date of valuation assuming:
   (a) a willing seller;
   (b) a reasonable period within which to negotiate the sale taking into account the nature of the property and the state of the market;
   (c) that values will remain static during that period;
   (d) that the property will be freely exposed on the open market; and
   (e) that no account is to be taken of any individual bid by a purchaser with a special interest.

1.2 The Institution stresses that if a valuer considers it appropriate to apply any qualifying words to 'Open Market Value', the meaning of those words should be

               discussed and agreed with the client before instructions
               are finally accepted. The Valuer should incorporate in
               his report the agreed meaning of the qualifying words.
      1.3 It is emphasized that this definition can in no way
               override the definition of market value which may have
               to be adopted for the purpose of valuations for under any
               statute.

Certain features of this definition are clearly concurrent with the general concept of 'open market value' as envisaged for taxation purposes. It is equally significant however that paragraph 1.3 recognizes that in particular areas the requirements of statute will be at variance with the RICS definition, and that the RICS definition cannot override any statutory definition.

Taking, first of all, the criteria by which such an open market value can be tested, it is clear that the value is to be determined without restriction, (such as on forced sale). Secondly by specifying the 'best price' it eschews any unduly conservative opinion as to the value. It is envisaging the best bid which might be anticipated in open tender (but subject to the specific exclusion of a purchaser with a special interest), moderated with the expectation that the best price is in all respects a 'reasonable' one.

The definition is equally clear in its requirement that the value estimated is to be by reference to a specific date. A valuation which fails to identify the specific interest in land *and* the specific date at which the worth of that interest is estimated would be quite useless.

The definition as it stands, therefore, is broadly harmonious with the intention of legislation in respect of valuation for taxation as this appears to have been interpreted. However, the assumptions which underlie the definition do not sit comfortably with valuation for taxation in their entirety. For example the basic concept for tax purposes appears to assume both a willing seller and a willing buyer. The RICS definition by including 'willing seller' thereby implicitly excludes a willing buyer. More importantly the RICS rules do not assume 'an individual bid by a special purchaser'.

This is not entirely out of line with the basic concept deriving from taxation legislation. That concept admits the special purchaser but appears to accept that the actual valuation will lie somewhere between a general market level and an exceptional bid.

This basic concept does not sit comfortably with parallel legislative provisions in the law of compensation for compulsory acquisition. The principles governing valuation contained in the Land Compensation Act 1961[17] contain a rule which appears to be designed to exclude special purchasers.

The special suitability or adaptability of the land for any purpose shall not be taken into account if that purpose is a purpose to which it could be applied only in pursuance of statutory powers or for which there is no market apart from the special needs of a particular purchaser or the requirements of any authority possessing compulsory purchase powers.

However, it has to be said that this reasonably clear statutory provision has been eroded by case law. In IRC v. Clay and Buchanan[18] it was held that a 'willing seller' is not to be assumed as one who would sell for any price (given that the land only had value to one apparently excluded purchaser). In VJC Raja v. Vizagapatam[19] a similar matter went to the Privy Council which held that where land was not for all practical purposes worthless then the existence of one only special purchaser could not be ignored and that a 'reasonable price' could be expected.

Later Lands Tribunal cases appear to confirm this approach and indeed have even gone as far as inventing intermediate speculators in order to remove 'special purchase' status from acquiring authorities.[20,21]

CONCLUSIONS

Valuation for taxation purposes can be difficult so far as the determination of market value is concerned. The establishment of market value in circumstances where there is an arm's length negotiation between unconnected persons for consideration in cash will be straightforward. But in almost any other set of circumstances presented to the valuer assumptions will have to be made. It is here that the difficulties will present themselves.

The approach to valuation which should commend itself in these circumstances will be to adopt the RICS guidelines as the best exposition of modern practice and capable of being scrutinized upon publication. Depending upon the specific taxation legislation which requires the valuation then the detailed valuation requirements of that legislation must be observed. Given, however, that valuation for taxation is not a legal concept, the judgement of the valuer should not be lightly usurped where there is any scope for interpretation of the statutory wording.

ALLOWING FOR THE EFFECT OF TAX IN VALUATION

As will have been realized from a reading of Chapter 3 almost all income is taxed and income from property is no exception.[22] The most important

Schedules for income from property are A and D Cases I and VI. Where there is a difference between income or profits gross of tax and the amount actually received, then companies and individuals will be concerned to know what their actual net of tax receipts are. It will only be possible to calculate the exact return on capital to an individual investor by using a figure which represents net income.

Since there will be a variation in income between owners of property giving variation in net-of-tax receipts from property, then a valuation of a given interest in land will have to take the individual circumstances into account in order to enable that person or company to consider the actual rate of return against opportunity cost or to consider whether the purchase of land or property is justified at any suggested value. Some consumers of and investors in property will not be liable to tax in certain circumstances. These will include charities, most obviously, but also major investors in commercial property investment markets such as superannuation and pension funds who will not pay tax on investment income to avoid double taxation of that income in the hands of those entitled to contractual pension rights, and who are paying for these rights out of taxable income during their working lives.

When a net of tax valuation has to be prepared of an asset for a particular person or company there are relatively few problems to be dealt with. However, given the heterogeneous nature of the owners and investors of property, it can be difficult to defend a net-of-tax valuation as one likely to meet the tests of acceptability for 'market-value'. The valuation so produced will naturally tend in the direction of the expectation of the generality of potential bidders for an asset, but will still be different from the actual net-of-tax valuation which each of these potential bidders might consider to be the figure (or less) at which they may be 'willing and able to purchase' the asset.

For this reason most analyses of sales and purchases of interests in land will be made on the basis of income received gross, i.e., before deduction of tax. Only in this way can any meaningful comparison be made since this is the only means of neutralizing the effect of tax.

Indeed in many circumstances this will not leave any residual problem. The recent reduction in Income Tax rate bands and levels means that net-of-tax income levels will be less widely different from when very high marginal rates of tax (of up to 98p in the £) obtained in the United Kingdom. For asset valuation purposes therefore potential buyers or sellers could be considered in the aggregate.

Freeholds

Certainly when income from a property asset is received in perpetuity than a valuation of that asset on either a gross-of-tax or net-of-tax bases will give the same result. This can be illustrated as follows.

Example 9.1

Consider a property asset which produces an income of £50 000 per annum and where a taxpayer's marginal rate of tax is 25%.

Gross-of-tax valuation	Rental income	£50 000
	YP in perpetuity @ 10%	10
	Capital value of interest in land =	£500 000
Net-of-tax valuation	Rental income	£50 000
less	income tax @ 25p	12 500
		37 500
	YP in perpetuity @ 7.5%	13 333
	Capital value of interest in land =	£500 000

Since the income has been reduced by taxation then it is capitalized at a net-of-tax rate giving a higher multiplication factor and the same valuation result.

Thus a gross-of-tax or net-of-tax approach to valuation will always produce the same result – but only where the income is both unvarying and perpetual.[23] These are rare conditions to be found when interests in land are being valued.

Freeholds with varying income

Conditions where the income varies will be more commonly found, and where such conditions exist then differences will arise in the valuation of the same asset as between gross- and net-of-tax valuations, and also as between net-of-tax valuations depending upon the rate of tax which is applied. These cases are illustrated as follows.

Example 9.2

1. Consider a property asset which produces an income of £50 000 per annum. On review in five years' time an income of £75 000 is expected. The gross-of-tax rate of return on capital is 10%.

Gross-of-tax valuation	Rental income	£50 000
	YP for 5 years @ 10%	3.7908
		189 540
	Reversion (in perpetuity) to rental income	£75 000

	YP in perpetuity @ 10% deferred 5 years	6.20921	465 691
	Capital value of interest in land		655 231

2. Assuming the same set of circumstances but making an allowance for taxation of the income at 25% then the value of the interest in land will be calculated as follows.

Net-of-tax valuation	Rental income	£50 000	
	less income tax @ 25p	12 500	
		37 500	
	YP for 5 years @ 7.5%	4.0459	
			151 721
	Reversion (in perpetuity) to rental income	£75 000	
	less income tax @ 25p	18 750	
		56 250	
	YP in perpetuity @ 7.5% deferred 5 years	9.28745	
			522 419
	Capital value of interest in land		674 140

3. Assuming the same set of circumstances but making an allowance for taxation of the income at 40%, then the value of the interest in land will be calculated as follows.

Net-of-tax valuation	Rental income	£50 000	
	less income tax @ 40p	20 000	
		30 000	
	YP for 5 years @ 6%	4.2124	
			126 372
	Reversion (in perpetuity) to rental income	£75 000	
	less income tax @ 40p	30 000	
		45 000	
	YP in perpetuity @ 6% deferred 5 years	12.45430	
			560 443
	Capital value of interest in land		686 815

As can be seen, a net-of-tax valuation of an asset with an anticipated higher future income will be more than a gross-of-tax valuation of the same asset. Similarly, higher rates of tax will cause net-of-tax valuations to be greater.

The underlying reason for this effect is the fact that any incidence of tax and any increase in the level of tax will mean that the future income is discounted at lower rates of interest with higher consequential capitalized amounts. (Relatively a lower value is being placed on the present income.)

So that where there is likely to be differences in the tax accountability of potential bidders from the same asset, then only a gross-of-tax valuation is likely to be useful as common to all and further usable for comparable purposes. For this reason market values of interest in land are usually prepared in this way. However, since it is only the net return on capital which is of interest to an investor in the final analysis, then a net-of-tax valuation will be necessary so that the true worth of the asset is revealed.

Leaseholds

In the case of leasehold valuations there are more fundamental problems for the valuer. As valuers will be aware it is conventional practice to allow for an annual contribution to a sinking fund designed to accumulate to the amount of the original cost of the asset. In this way the amount of the capital is preserved in nominal if not in real terms. Of course, the monies allocated to the sinking fund are extracted from the rental income so that the net receivable amount is diminished. Also the periodic payments in respect of the sinking fund have to be made out of income which may be taxed as profit in the hands of the owner of the asset.

Where tax is due on rental profits then certain adjustments have to be made to arrive at a net-of-tax estimate of the net worth of the leasehold interest to the tax payer. Since tax is suffered equally by the portion of income retained for the purchase of an annuity and the portion of income released as net profit then the profit retained has to be further reduced in order to compensate the income retained for the sinking fund for the amount lost to tax at the appropriate rate. To do this the annual sinking fund contributions are increased in the ratio of income before tax to income after tax by multiplying the annual sinking fund by the factor

$$\frac{100}{100 - T}$$

where T equates the rate of tax expressed as a percentage.

Hence, the Years' Purchase formula for a net-of-tax leasehold becomes

$$YP = \cfrac{1}{i + \left(SF \times \cfrac{100}{100 - T}\right)}$$

Where, i = the remunerative rate of interest as a decimal percentage.
SF = the annual amount of the sinking fund necessary to recoup
£1 during the term of the lease at an appropriate rate of
interest (net of tax).
T = percentage rate of tax.

Supposing the leasehold interest were for fifteen years, at a remunerative rate of 10% per annum and a sinking fund rate of 3% (net) per annum with the tax payer then liable to tax at 35p in the £, the Years' Purchase factor used for capitalizing the rental income would be derived as follows:

$$YP = \cfrac{1}{0.1 + (SF \text{ to accumulate £1 in 15 years @ 3\%}) \times \cfrac{100}{100 - 35}}$$

$$= \cfrac{1}{0.1 + (0.0537666 \times 1.5384615)}$$

$$= \cfrac{1}{0.1827178}$$

$$= 5.4729201$$

This YP factor is known as the Years' Purchase for fifteen years at 10% and 3% adjusted for tax at 35p in the £.

Example 9.3

Calculate the net-of-tax worth of a leasehold interest in land where the profit rent is £5000 per annum receivable over the term of the lease which is nine years. The required yield is 6% and a sinking fund can be established at 3%. Tax is payable at 37%.

Net-of-tax valuation	Profit rent	£5000
	YP for 9 years @ 6% and 3% (tax @ 37%)	4.6244
	Capital value of interest in land	= £23 122

The determination of the YP multiplication factor can be made more explicit as follows:

YP for 9 years @ 6% and 3%

$$= \frac{1}{0.06 + (\text{SF to accumulate £1 in 9 years @ 3%})} \times \frac{100}{100 - 37}$$

$$= \frac{1}{0.06 + (0.0984339 \times 1.5873016)}$$

$$= \frac{1}{0.2162442}$$

$$= \underline{4.6244}$$

Observations on valuations
The illustrations provided in this chapter reflect the conventional approach to valuation and a number of adjustments may have to be made in particular cases. Clearly each calculation can be carried out using an electronic calculator with logarithms and exponential functions. The calculations will more easily be resolved if a calculator with financial functions is used. In either case explicit account can be taken of each of the variables in any equation. Thus if a rent is paid quarterly in advance or half-yearly in arrear instead of the annual payment in arrear allowed for in the examples, then the necessary adjustments can easily be made. Students of valuation will appreciate the flexibility which such technology affords, but many practitioners only prepare valuations using valuation tables. This means that the more limited approach overhangs the market. In any case actual differences in valuations are often only marginal. Readers should be aware, however, that more precise answers to valuation problems can be obtained.

Similarly, there are variables which may be important in an actual valuation which have not been dealt with here. These would include expectations of income growth rates and depreciation of the value of the asset, for example. Instead an 'all-risks' yield approach has been used, which is again the convention in the market place.

Within the simple illustrations provided of valuations for taxation purposes, enigmas remain which may need clarification even if they cannot always be justified. For example, the perceived need for a sinking fund may be observed more in the breach than in practice. More particularly the appropriate rate to choose to accumulate the (annual) periodic payment is contentious. There appears to be little evidence of such facilities being quoted in the market place and the level of interest used appears to be conservative. Also the convention is to quote the interest rate on sinking fund policies on a net-of-tax basis.[24] The convention has been adopted in these illustrations so that a 3% rate of return net-of-tax will be the equivalent of a 4% rate of return before deduction of Income Tax at 25p in the £.

It should also be noted that no adjustment for tax has been made to the remunerative rate of interest. Hence no deduction has been made for tax to reduce gross income to spendable income. Again this is simply a convention which allows for comparison between yields where the actual effective marginal rate of tax may be in any case difficult to determine.

Investment properties

It has been mentioned a number of times that since yields considered gross-of-tax are most suitable for comparison between types of property assets and between interests in land and other income-producing assets then gross-of-tax yields are conventionally used in land valuation. It has equally been re-iterated that the effect of tax is important and that it is the net-of-tax valuation which is important when making an individual or corporate decision on the disposition of property assets.

Clearly there will be occasions where consideration of the net-of-tax position will be of crucial importance and this is particularly true of property investments. When there were higher rates of tax and substantial differences between income and capital tax rates then a net-of-tax approach could be particularly revealing. With reduction in tax rates and increasing harmonization between income and capital tax rates then tax planning decisions may be more subtle.

In circumstances where decisions have to be made between alternative investment decisions then the net-of-tax appraisal will be most useful. A classic example is the one where an investor has to make a decision about property which he already owns. The following example provides an illustration.

Example 9.4

Mr A owns a property which he purchased in 1972 for £25 000. The value at March 1982 was £50 000. In May 1988 he received an offer to purchase from the sitting tenant in the sum of £98 000. The net annual income from the property, after outgoings but before taxation, is £11 000. Mr A's opportunity cost for investments of comparable risk is 13% and his marginal rate of tax is 37.5%. Should he accept the offer?

Present investment income

		Net rental income	£11 000
less		income tax @ 37.5%	4 125
		Net-of-tax income	£ 6 875

Liability to Capital Gains Tax

Proceeds on disposal (net)	say	£95 000
Value at March 1982		50 000
Unindexed gain		£45 000
Indexation allowance		

$$\frac{RD - RI}{RD} = \frac{106.2 - 79.4}{79.4} = 0.338$$

50 000 × 0.338		16 877
Chargeable gain on disposal	=	28 123
Capital Gains Tax payable at tax payer's rate of 37.5%		10 546

Alternative investment income

Capital available for re-investment		
= £95 000 − 10 546	=	84 453
Gross income receivable @ 13%	=	10 798
Less Income Tax at 37.5%	=	4 117
Net-of-tax income		£ 6 860

This analysis demonstrates that Mr A would be slightly worse off net-of-tax if he were to sell his freehold interest to the sitting tenant for a sum of £98 000 and re-invest in an asset of comparable risk. If the sale had been concluded in fiscal year 1987/88 the rate of tax would have been lower (30%) and the chargeable gain would have been lower. If his marginal rate of Income Tax was the same as now the sale would have been justified. As it is he will require proceeds of disposal greater than £98 000 to justify the decision to sell.

Example 9.5

Mr A has the opportunity to purchase a short leasehold which will produce a net income of £2500 p.a. over its remaining term of six years for a sum of £7000. His rate of tax is 35%. Comparable leaseholds have a remunerative rate of 11%.

Valuation of leasehold interest

Profit rent	=	£2500
YP for 6 years @ 11% and 3% (tax @ 35%)		2.8749
		7187

On the face of it the purchase price may be fair as it is slightly undervalued, until the investor considers the incidence of tax.

Net-of-tax valuation

Profit rent		=	£2500
less Income tax @ 35%			875
	Free income		1625
YP for 6 years @ 11% × 3% (tax @ 35%)			2.8749
			4672

As can be seen, the incidence of tax has a highly corrosive effect on Mr A's spendable income so that a purchase price of £7000 would require a major sacrifice in the remunerative rate of return.

This may be bad enough, but the net-of-tax position will deteriorate even further if the sinking fund is adjusted to allow for Income Tax liability.

The annual sinking fund required to amount to £7000 at the end of six years is £1082, after tax. If this figure were grossed up to take account of Mr A's marginal rate then the annual sinking fund required would be

$$£1082 \times \frac{100}{100 - 35} = £1664$$

Accordingly his net-of-tax income from the investment will become

Rack rent	=	£2500
less gross sinking fund	=	1664
Gross income	=	836
less tax @ 35%	=	293
Net of tax income	=	543

This will provide Mr A with a net yield after tax of 7.76% on his investment of £7000. In comparison a gross fund such as a pension fund, with no liability to tax on investment income, will enjoy an IRR of 27.34% on the net rack rental income at the same purchase price.

With short leaseholds, where the sinking fund has greater impact the shorter the term, and where differing rates of tax will have relatively significant affects on sinking fund costs then the notion of 'market value' may be difficult to sustain. At best only an average rate of tax calculation for investors grouped according to their tax personae may suffice. For each individual investor the net-of-tax return will be the only figure which s/he should use for the purpose of bidding for the asset. Among bidders for short leasehold the more aggressive will be gross funds and it is this fact which explains the noticeable activity of superannuation and pension funds in recent years in this sector of the property market.

NOTES: CHAPTER NINE – LAND VALUATION

1. (1914) 3 K.B. 466.
2. *IRC* v. *Marrs Trustees* [1906] 44 S.L.R. 647.
3. *Buccleuch (Duke)* v. *IRC* [1918] 2 K.B. 735.
4. [1965] 3 All E.R. 458; [1967] 2 W.L.R. 207.
5. This may be compared with the practice adopted in the market place whereby a bidder for a development site will deduct a percentage from his estimate of market value to adjust for disposal costs when calculating his bid on the basis of a residual valuation.
6. (1906) 44 S.L.R. 647.
7. (1915) S.C. 449.
8. [1927] 1 C.H.D. 313.
9. [1970] C.H.D. 318.
10. *VJC Raja* v. *Vizagapatam* [1932] A.C. 302.
11. *Finlay's Trustees* v. *IRC* [1938] 22 A.T.C. 437.
12. *Salomon* v. *Customs and Excise Commissioners* [1966] 2 A.E.N. 340.
13. *Buccleuch (Duke)* v. *IRC* [1967] 1 A.C. 506.
14. *IRC* v. *Marrs Trustees*, Supra.
15. For example the Stock Exchange 'Yellow Book' (for admission of securities to listing) requires that current accounting conventions be observed and that in the case of property valuation due regard be had to the guidelines issues by the RICS of which it approves, and which define 'open market value'.
16. RICS (1989), *Guidance Notes on the Valuation of Assets* (2nd Ed.) (as amended), RICS, London, p. 230, Guidance Note No. G.N. 22. See also, RICS (1989), *Manual of Valuation Guidance Notes*, RICS, London, Appendix 2, p. 1.
17. See 5(3) in Scotland the Land Compensation (Scotland) Act 1963 Sec.12(3).
18. [1914] 3 K.B. 466.
19. *VJC Raja* v. *Vizagapatam* [1932] A.C. 302.
20. *Rathgar* v. *Haringey London Borough* [1978] 248 E.G. 693.
21. Blandrent Investment Developments v. British Gas Corporation [1978] 252 E.G. 267.
22. It should, however, be remembered that certain sources of income are exempt from tax including certain Gilt Edged Stocks and National Savings Certificates. Other income, such as that from bank and building society savings accounts are paid net of tax at an aggregate rate and not at the individual's personal rate, if received prior to 6 April 1991.
23. The gross-of-tax and net-of-tax valuations of leaseholds will be identical (even where the income is not perpetual) but where income is unvarying and 'dual-rate' method is used (see infra).
24. This convention is adopted in Parry's Valuation Tables. It is assumed that tax is paid by the financial institution issuing the policy before payment is made to the policy holder.

Chapter Ten

TAXABLE SITUATIONS
AND TAX PLANNING

INTRODUCTION

As has been indicated previously the taxation system is comprehensive and is therefore not concerned with single and discrete events. An interest in land may be sold in its entirety or in part, a lesser interest such as a lease may be created, the interest may be sold and leased back to the original vendor as a condition of sale, land may traded as a commodity, assembled and developed, it may serve as an investment medium or may be consumed by an occupier. These possibilities for the use and transaction of land create potential liability for the incidence of taxation and such taxable situations may cause the interaction of two or more different forms of taxation.

It is the intention of this chapter to provide a brief and selective insight into the way differing events in respect of interests in land can bring about taxable situations and to highlight any concomitant land valuation considerations.

Clearly owners of land and their advisors will also wish to act prudently in the conduct of any taxable situation in order to minimize any liability to tax. If alternative courses of action afford the same acceptable outcome, but if one of these courses can be more tax-efficient than the others, then is it not sensible to examine alternatives and their fiscal results with a view to choosing a preferred means of dealing with an event involving an interest in land? Similarly, the effluxion of time can bring taxable events into play which might otherwise have been avoided or ameliorated. Tax planning is therefore an important means of tax minimization and a necessary activity when considering the management of property.

A few taxable situations and tax planning measures are illustrated here in order to provide some insight into the tax implications of land transactions. It is important for the reader to appreciate that these illustrations are not comprehensive and that the tax planning pointers are ones which may be available in the circumstances but are not necessarily the best ones, or even the correct ones in the detailed circumstances of any particular case. Tax planning is most easily approached by reference to generic groups of taxpayers such as property dealers, property investors, rural landowners and so on. Such an approach would demand a degree of detail which would far exceed the purposes and intentions of this text.

Readers should not therefore rely on the information given in this chapter for dealing with actual taxable situations, or with real life tax planning measures. The contents of the chapter are merely intended to give a flavour of the interaction of taxes and an indication of possible avoidance measures or available reliefs.

INTERACTION OF TAXES

Another cautionary prefatory comment is necessary. Any given taxable situation can result in liability to more than one tax. The method by which legislation permits the calculation of taxes based on their inter-relationship may be important and could affect a decision on the disposition of the interest in land.

For example, a disposal could have brought about a simultaneous DLT and CGT event. Here the rule would be that DLT will be payable in full (ignoring CGT liability). However, the actual amount of chargeable realized development value will then be deducted from the actual chargeable gains for CGT (DLTA 1976 Sch.6 para.1(1) and (2)). But a DLT disposal may antidate or postdate a CGT disposal when different rules will apply.

Again where a disposal of an interest in land is a gift *inter-vivos* then there may be liability to both IT and CGT, but where the transfer is on death then the tax payer's estate will be valued at that date. Although there may be a gain in the value of the assets the estate will be liable to IT only and not CGT. The position will also be affected by the taxation provisions made with a gift of an interest in land. If tax is to be borne by the transferors then both IT and CGT will be payable (ITA 1984, s.5(4)) but if the transferee is to bear tax than the amount of CGT, if actually paid, will be deducted from the value of the interest in land transferred (ITA 1984 s.165(1) and (2)).

The means by which provision is made for the interaction of taxes varies between and among different taxes and different situations. The legislative provisions for the interaction of taxes is complicated and is yet another matter which is beyond the scope of the simple description of the tax system contained in this text.

Readers should be aware of the importance of the interaction of taxes when considering the potential impact of taxable situations and tax planning measures. Reference to more detailed texts is recommended should these considerations be important.

LEASES

The extent of tax liability will depend upon the nature of the lease. A lesser interest may be created out of a freehold by way of the grant of a head-lease. A subsequent grant of a sub-lease will give rise to different taxation circumstances as will an assignation of a lease or the disposal of a reversion. Some of these differing circumstances will be examined here.

Grant of a head-lease

Tax may be liable under the following headings

1. *Development Land Tax.* If the granting of the lease releases development value then there will be liability to DLT on 'the landlord's rental rights' as a part disposal (DLTA 1976 Sch.2). As the reader will be aware the calculation, including the valuations involved, can be onerous. Fortunately, however, such liability will only apply to events which occurred before 19 March 1985.

2. *Income Tax.* Rental profits will be liable to Income Tax under Schedule A (ICTA 1988 s.21(1)) on an annual basis throughout the duration of the lease.

    If the lease is for a term of fifty years or less and the lease is paid for partly, or wholly, by way of a premium then there will be a liability to tax under Schedule A upon the granting of the lease. In these circumstances premiums are treated as income in the form of rent (ICTA 1988 ss.34–39 and Sch.2, FA 1988 s.73).

3. *Capital Gains Tax.* If the granting of the lease requires rental payment only (and provided that such rent is at full market value) then the liability to taxation will be solely to Income Tax and Capital Gains Tax will not apply.

    In circumstances other than these there could be liability to Capital Gains Tax under the part disposal rules (CGTA 1979 s.19(2)). Where a lease is granted subject to the payment of a premium (or other considerations for the grant of the lease), then there will be a part disposal of the asset (CGTA 1979 Sch.3 para.2(1)).

4. *Inheritance Tax.* If the granting of a head-lease results in a transfer of value by which the value of the tax payer's remaining estate is diminished then there is *prima facie* liability to Inheritance Tax (ITA

1984 s.1)). The general case will be where property is leased at full market value, when a charge to Inheritance Tax will not arise. In certain instances this is confirmed by statute. Thus the grant of a tenancy of agricultural property (for agricultural use) is specifically not a transfer of value if a market rent is paid (ITA 1984 s.16). Again the granting of a lease at full market rent is not considered to be an associated operation[1] provided that the operation associated with the granting of the lease is carried out more than three years after the commencement of the lease (ITA 1984 s.268(2)).

Remembering, however, that the test for liability to IT is reduction in the value of the transferor's estate by reason of the disposition, then circumstances can arise where a reduction of the value of the estate is brought about as a result of the granting of a head-lease. Suppose, for example, that an owner of domestic property granted a lease to a tenant at a full market rental it is possible to imagine that in these circumstances the value of the landlord's interest in land might be diminished. The value of the property subject to the lease may be less than the vacant possession value. The value of the owner's estate has therefore been diminished and liability to IT may arise.

Grant of a sub-lease

The grant of a sub-lease is similar to the grant of a head-lease in that a lesser interest is created and a part disposal occurs. Broadly, therefore, the same taxes will apply as for the grant of a head-lease.

1. *Development Land Tax*. In the (increasingly unlikely) circumstances where a case requires to be determined for a taxable event occurring before March 1985 where development value has been released as a consequence of the granting of a sub-lease, then special considerations will arise in respect of sub-leases. The position will depend upon whether the sub-lease is a short or a long lease, the determining period of time being the usual term of fifty years. In each case a relatively large number of valuations will be required.[2] Reference to a specialist text is therefore recommended should DLT apply.

2. *Income Tax*. As previously, for the grant of a head-lease there will be liability to tax under Schedule A if the sub-lease is granted for a term less than fifty years and with the payment of a premium (or equivalent consideration) for that part of the premium treated as income, and for the periodic rent received.

   If the lessor paid a premium at the time he was granted a head-lease and if he is receiving a premium upon the granting of a sub-lease then he can expect relief on the latter. As a head-lesee may deduct the head-rent from the sub-tenancy rents when calculating profits, so he can logically deduct the payment of the superior premium from the inferior (ICTA 1988, s.37).

3. *Capital Gains Tax.* The taxation provisions are similar to those applicable to the grant of a head-lease. If a premium is paid by the sub-lesee then the amount may be some or all of the acquisition cost paid and on which the superior landlord may have made a gain. If the grantee of the sub-lease is a tenant under a lease with a term of fifty years or less then the sub-tenant, when being considered for liability under CGT, can amortize the premium over the duration of the lease.

4. *Inheritance Tax.* See under grant of a head-lease, supra.

Assignment of a lease

In the event that a lease is assigned from one party to another, at open market value, there will be potential liability to DLT and CGT. The calculation for DLT purposes follows the general rules for the transfer of an interest in land, and because of remoteness the detailed application of the method is not illustrated in this text.

In the event that a lease is assigned from one party to another at less than market value there may be liability to additional Income Tax. There could also be liability to IT when the general rules for calculation of liability will apply.

1. *Income Tax.* The assignment of a lease is normally for a capital sum and no liability to Income Tax will arise other than in the (unusual) circumstances where a trader may be liable under Schedule D if the sale has been realized to make a profit.

    However, in circumstances where a lease is assigned at under value there may be liability to Income Tax under Schedule A (ICTA 1988 s.35(1)). This legislative provision requires the identification of the amount forgone, i.e., the amount of any premium that could have been charged on the grant of the lease by the original landlord.

2. *Capital Gains Tax.* Where a lease is assigned which has a life greater than fifty years then the normal CGT rules apply in the same way as for the sale of any interest in land which is an asset for CGT purposes.

    Leases of fifty years duration or less at the date of the taxable event are short leases so far as CGT is concerned and are treated as wasting assets (CGTA 1979, Sch.3. para.1(3)). The rules specific to leases as wasting assets are then deployed.

SALE AND LEASEBACK

Introduction

Sale and leaseback transactions have become an established means of raising finance. They invoke a simple device which is that the disposition of a major interest in land is sold conditional upon the purchaser immediately granting a lease to the vendor and for a mutually agreeable term. The device thus ensures that the vendor can raise capital while at the same time retaining possession of the land.

The device has its genesis in the era of post-war credit restrictions and proved to be an effective tool for property developers. Its robustness as a means of raising finance can be attested to by the way in which it has continued to flourish in a climate of increasing financial deregulation. Partly its success may be attributed to the breadth of its application and to its flexibility. It has been widely used in the purchase of single farms by financial institutions while being imaginatively developed in a variety of commercial development packages in the City of London and elsewhere.

The simple form of sale and leaseback is outright sale of the total interest at open market value, with a leaseback by the vendor at full rack-rented value. Of course, the terms may be varied. Part only of the interest may be sold. The sale may be for an amount less than open market value with a commensurate reduction in rent payable from full rack rental value – or vice versa. The interest in land may be a lease and not a freehold.

There are specific legislative provisions covering sale and leaseback arrangements entered into by the holder of existing leases and which take place after 21 July 1971. These provisions are operative when a leaseholder with a lease of more than fifty years to run sells her/his interest for a capital sum and then takes a lease back for a term not exceeding fifteen years. They also operate where no capital sum is paid but where a rent in excess of full market rental value is agreed in the early part of the lease for a period not exceeding fifteen years (ICTA 1988 s.36(3)).

Development Land Tax

If development value could be realized then the transaction could attract DLT if it took place before 19 March 1985 (and, after 31 July 1976).

Normally the transaction will have been a part disposal. (The transaction is effectively a sale of an interest which is conditional upon a retention of part of that interest through the grant of a lease.) The DLT part disposal rules will, therefore, usually apply.

Because of remoteness the detailed implications of liability to DLT are not pursued here. However, it should be pointed out that if DLT applies then there may be *prima facie* laibility to double taxation.

Income Tax

If the interest in land which is sold is actually trading stock or has otherwise been bought in order to secure a profit then the proceeds on disposal will be subject to Income Tax in the usual way. Otherwise there will normally be no liability to Income Tax.

However, Income or Corporation Tax may be payable on part of the proceeds from a sale and leaseback transaction. Substantively this will depend upon the length of the lease with a greater tax liability accruing the shorter the term of the lease agreed.

In the circumstances described in the introduction (supra) where a lease is granted for a period of fifty years or less, part of the premium is treated as income.

Capital Gains Tax

Given the nature of a sale and leaseback transaction then the actual circumstances of the case will be necessary to decide whether there has been a total or part disposal of an interest in land. In the circumstance where a gain has been made there will be liability to Capital Gains Tax and the normal CGT rules will apply.

DEALING AND DEVELOPING

Introduction

It is not usually difficult to identify whether a profit has been made on a land transaction. It may be very much more difficult to attribute the correct motivation or intention to the parties involved and/or to categorize the heading(s) under which tax will be liable.

We have seen (in Chapter 3) that transactions in land can come under a variety of headings. On the sale of an asset tax could be payable under Schedule D, Case I (as the profits of a trade or business) or Schedule D, Case VI (as the profit from development gains)[3]. Or it may be that neither of these categories is correct and that the charge to tax may be under Capital Gains Tax.[4]

In order to determine the correct taxation provisions, then certain questions have to be addressed. Is the tax payer trading in land or realizing an investment? If the tax payer is trading in land then in what manner is his profit realized?

Identification of trading

The legislation defines a 'trade' as including 'every trade, manufacture, adventure or concern in the nature of trade' (ICTA, 1988 s.832 (1)). Unfortunately, this definition is not as helpful as it might appear at first sight and the question of whether a transaction in land is 'trade' has to be answered with reference to the facts of the matter. Not surprisingly there is now a great deal of case law which assists with dealing with this problem by the application of a variety of litmus tests.

Capital v. Income

As we have seen (in Chapter 2) the distinction between capital and income is uncertainly treated in the UK tax legislation. Equally, the distinctions between acquiring land for trading or investment are fine. But the category into which a transaction is placed will be important, and the tax payer may be able to do much to influence his legal obligations to pay tax and therefore the amount actually paid.

The basic distinction which the law makes between capital and income is clear. Income is perceived as a flow of money from a permanent (or reasonably permanent) asset. Hence a sum of money held in a bank account is capital while the interest received by the account holder is income. A shop property is capital while the rent received from it is income. Normally the taking of income from such assets does not cause any diminution in their value. The analogy often referred to is the fruit-bearing tree where the fruit is the income and the tree is the capital. The tree endures and produces recurrent crops of fruit in season.

The law provides that the flow of money from an asset should be taxed as income and that any increment in the value of the asset from which the income is derived should be taxed as capital.

But this simple state of affairs can be blurred in real life. Transactions in land can arise in very diverse circumstances and for widely different motives. It is essential therefore that the circumstances of each case are examined carefully and the facts established as clearly as possible.

The 'Badges of Trade'

Case law has suggested that certain factors are important when considering whether or not trade is being conducted in a land transaction and whether therefore there has been a realization of trading profit, or alternatively, capital gain. The main factor will be the actual motive or intention of the tax payer. Collectively these factors are colloquially known as badges of trade and some of these are individually considered below. A more comprehensive list of factors may be found in Marson v. Morton [1986].[5]

1. Intentions of the tax payer

Clearly there will be occasions where the intention of the tax payer was obscure or quite unknown. (As a matter of good tax practice, therefore, it is prudent for a tax payer to keep a record of what his intentions were, if any, when buying, selling or otherwise transacting interests in land.)

It may be that a person liable to taxation may have had no motive or intention at all when s/he acquired the property. The property may be settled or inherited, so that a subsequent realization of proceeds would not be considered *prima facie* a profit from trade.[6]

Otherwise it is a question of determining what the actual objectives of the tax payer were at the time of the transaction(s). If the original intention of the tax payer was not to sell then a subsequent intention to sell will not constitute a trading operation even when that subsequent intention includes the application for and obtaining planning permission for the development of the property (Taylor v. Good).[7] Equally if the original intention of the tax payer was to resell the property at a profit then this will constitute an adventure in the nature of trade (Clark v. Follet).[8]

So that if there is evidence one way or the other as to the tax payer's actual intentions at the time of acquisition of the property, or at the commencement of a series of events, then this will be ruled upon when considering whether an adventure in the nature of trade is being pursued.

In the case of Johnson v. Heath [1970][9] an employee of a building company negotiated an enhanced sale price of land (£25 000) before he himself had purchased the land from the company for a sum of £15 000. It was held that the tax payer had a clear intention to sell the property before he purchased it and that the transaction was in the nature of trade and not an investment as the tax payer had claimed.

The fact that those involved in land transactions had trade knowledge, was held to be relevant in the case of Burrell, Webber, Magness, Austin and Austin v. Davis [1958][10] where it was held that a group of five persons who acquired property and resold it profitably after a short period of time had no intention of holding the property permanently and that the intention to make a profit out of a transaction could be partly attributed to the knowledge of two of the men who had previous dealings in property.

Establishing the intentions of the tax payer will still give difficulty where a change of intention occurs between the date of acquisition and the date of sale. It will need to be established whether the tax payer, when selling advantageously, is simply disposing of an asset in a prudent manner or whether s/he is now in pursuit of profits from trade.

The intentions of the taxpayer can be even more difficult to discern where the participation of others is involved. Situations have existed in the past where quite elaborate arrangements were made to avoid dealing transactions being so categorized. A leading case in this regard was Ransom v. Higgs[11] where a complicated scheme was devised whereby Mr Higgs and his family were beneficiaries through a trust of profits made via a series of transactions which he organized. He did not own the land involved but he organized the transactions. It was held that Mr Higgs had not carried out a trade but had simply arranged a trading profit for another.

This case may be said to have been a typical example of the statutory interpretation of tax law mentioned in Chapter 2. To deal with such situations legislation was introduced specifically 'to prevent avoidance of the tax by persons concerned with land or the development of land' (ICTA 1988 s.776 (1)).[12] S.776, which only applies to land, seeks to neutralize avoidance by attempting to identify a wide range of transactions and then charging gains of an apparently capital nature as liable to Income Tax. It is attempting to identify arrangements made to conceal the fact that the tax payer is essentially a dealer in property. For this reason it is wide-ranging although it leaves the Inland Revenue with scope for discretionary exclusion.

In the particular case of land therefore there is a major piece of anti-avoidance legislation. Where a transaction or a series of transactions is likely to bring s.776 into play then particular consideration should be given to its powers. The detailed pursuit of these would be inappropriate here. It should also be noted that many of the cases mentioned in this section of the text refer to dates before s.488 (ICTA 1970) and even before the introduction of capital gains (FA 1965), so that a simpler and more deterministic view was taken of trading. If there was no trade and no liability to Income Tax, then there might be no liability to tax at all. The decisions made in more contemporary cases such as Marson v. Morton[13] are therefore more salient.

It should be further remembered that for any transactions (not only land transactions) it will be possible to examine not only the incidence of taxation within one step of a series of transactions but also the structure of the whole transaction. Such an approach may be more effective in revealing operations which are an adventure in the nature of trade. (See Chapter 1, and the decisions in W. T. Ramsay v. IRC,[14] Burmah Oil Co. Ltd v. IRC,[15] Furniss v. Dawson[16] and Craven v. White[17]).

Since intentions may be difficult to identify regard may be had to other circumstantial evidence of the kind mentioned below.

2. History of similar transactions

If there has been a series of similar transactions then this can provide evidence of an intention to trade.[18] If the taxpayer has been involved in the same trade before then a subsequent operation which is sufficiently similar may be regarded as trading activity.[19]

Where previous trading has been established then subsequent transactions will be regarded as trading.[20] If a tax payer has traded in the past and if the property has formed part of his stock-in-trade then it will continue to be viewed as part of his stock-in-trade even if sold much later.[21] If stock is transferred from a trading account to a private account, then a change of intention may be admitted, and the subsequent sale of the property will not be seen as a trading operation.[22]

3. Tax payer's knowledge or expertise

If a taxpayer has particular knowledge or expertise because of his trade or profession and if this know-how is of direct value to the particular transaction then there may be a presumption that the transaction is in the pursuit of trade.[23] Many of the cases quoted here involve builders. Property developers and estate agents are much more likely to be categorized as being involved with a trading operation in land than an individual who has no such knowledge or expertise.[24].

4. Period of ownership

Normally it is an objective of trade to dispose of assets at a profit as soon as possible after acquisition and for this reason the period of ownership can be a useful indicator of whether a transaction in the way of trade is taking place. The longer the period of ownership then the less likely it is that a trading operation is occurring. Cases can be found where a relatively brief period of ownership can result in the transaction being categorized as trade,[25] and where a relatively lengthy period of ownership can result in the transaction not being categorized as trade.[26]

However, by itself the length of period of ownership will not necessarily provide conclusive evidence of an act of trade. If this is likely to give difficulty, say when a property is held purely for investment but may on occasion be sold to take advantage of market conditions or to arrange for portfolio adjustment then prudent steps should be taken to establish motivation at the time of purchase.

5. Improvement of the asset

Unlike period of ownership, the improvement or development of an asset will not necessarily give a primary indication of whether trade is being pursued. This aspect is fairly neutral in other than the basic situation where a property with development potential is acquired, planning permission obtained and then immediately sold.[27]

But an equally typical activity is the acquisition of property by an investor which is then improved in order to enhance its rental (investment) income. Equally land may be bought, developed and held pending a decision as to whether to sell it or for a trading profit or alternative hold it as an investment. In either of these cases an early sale could risk being caught under the provisions of Sec. 776 (ICTA 1988).

Conclusions

While these tests are by no means exhaustive (although any litmus test which shows a positive result on the tax payer's intention will be conclusive) enough has probably been said to convince the reader that it is important to distinguish between capital and income receipts and therefore between investment (or occupation) and an adventure in the nature or trade. Not only is this necessary in its own right but also because of the anti-avoidance provisions of tax legislation aimed exclusively at land and with the intention of exposing concealed dealing operations in land. Only after a transaction has been carefully and properly categorized having regard to the intention of the tax payer and the circumstances of evidence of the case, can an approach to the actual calculation of taxation liability be made and the necessary valuations identified.

TAXATION FOR DEALING AND DEVELOPING

Income Tax

If a company or a person is categorized as dealing in property as a trade, either because this is self evidently the case or because of the anti-avoidance procedures of Sec.776[28] (ICTA 1988), then the assessment of profit under Schedule D Case I will be based upon normal accounting practice. Profit will be determined by reference to the net figure remaining after allowable expenses are deducted from trading receipts for any given accounting period.

In the particular case of land then assets may be held for a length of time longer than the accounting period. Stock unsold at the end of the period may have its cost carried forward to the accounting period when it is disposed of. The original cost (or net realizable value, if lower) is then deducted as an expense occurring within that period.

For the most part normal accounting procedures regarding the calculation of profit and the valuation of stock will be expected and accepted by the Inland Revenue. But there are detailed points which may need further examination in practice, and special stock relief provisions which may apply (FA 1981 Sch.9 paras.9 and 17(1) and ICTA 1988 Sch.30 para.18).[29]

Allowable expenses will include monies spent on improvements to land and buildings and the cost of financing the acquisition of trading stock as will the cost of acquisition of assets and the reasonable costs of managing the properties and running the business. The tax payer's private expenses and other non-trading expenses are not allowable in the usual way (ICTA 1988 s.74).

Rental income from property assets which are held as trading stock will be assessed to Schedule A (See Chapter 3).

Development Land Tax

The incidence of DLT will be increasingly remote as time passes, but any outstanding cases on transactions prior to 19 March 1985 could include a liability to this tax. If the disposal of an asset, whether held as an investment or as stock in trade, causes the realization of development value, then DLT would apply.

However, the realization of development value and the realization of a trading profit are not necessarily co-terminous events. A chargeable event for DLT purposes can arise at times other than a disposal by way of trade as when a project of material development commences or when land and property, held as stock in trade is leased.

Where the events do run together, then provision is made for relief designed to avoid double taxation. Since a charge to DLT is usually taken first then the relief takes the form of a deduction of the DLT tax charge from the charge to income or corporation tax.

NOTES: CHAPTER TEN – TAXABLE SITUATIONS AND TAX PLANNING

1. Separate transactions, the value of the parts of which could be less than the whole, are identified in the ITA and are treated as associate operations which are then taken to have a single value.
2. Examples follow.
    1. The market value of the head-lease at the date of the sub-lease.
    2. The value of the landlord's rental rights under the head-lease at the time of the grant of the head-lease.
    3. The value of the landlord's rental rights under the head-lease at the time of the grant of the sub-lease.
    4. The value of the landlord's rental rights under the sub-lease at the time of the grant of the sub-lease.
    5. The CUV of the head-lease on these occasions; when it was granted, prior to the sub-lease and subsequent to the sub-lease.

3. Rental receipts and premiums taxed as income are excluded from trading profits and are taxed under Schedule A.
4. In addition there could be liability to DLT if the sale took place between 31 July 1976 and 18 March 1985.
5. S.T.C. 463.
6. Note, however, the decision *Balgownie Land Trust Ltd* v. *IRC* (1929) 14 T.C. 684, where the trustees of an estate formed a company out of which dispositions were made. Since the company's articles of association declared its intention to deal, then the dispositions were held to be in the nature of trade (see also *Pilkington* v. *Randall* (1965) 42 T.C. 622).
7. [1974] 1 All E.R. 1137.
8. [1973] S.T.C. 240.
9. 3 All E.R. 915.
10. 38 T.C. 307.
11. [1974] 3 All E.R. 949; S.T.C. 539; T.R. 281 HL.
12. Originally enacted in the Finance Act 1969 and later in the ITCA 1970 s.488.
13. [1986] S.T.C. 463.
14. [1981] All E.R. 865.
15. [1980] S.T.C. 731.
16. [1984] A.C. 474.
17. [1988] T.C. 476.
18. *Pickford* v. *Quirke* [1927] 13 T.C. 251.
19. *MacMahon and MacMahon* v. *IRC* [1951] 32 T.C. 311.
20. *Page* v. *Pogson* [1954] 35 T.C. 545.
21. *Oliver* v. *Farnsworth* [1956] 37 TC 51; 35 A.T.C. 410.
22. *Harvey* v. *Caulcott* [1952] 33 T.C. 159; 31 A.T.C. 90.
23. *IRC* v. *Fraser* [1942] 24 T.C. 498.
24. *Burrell, Webber, Magness, Austin and Austin* v. *Davis* [1958] 38 T.C. 307 and *Emro Investments Ltd* v. *Aller* [1954] 35 T.C. 305.
25. *Turner* v. *Last* [1965] 42 T.C. 518 and *Johnson* v. *Heath* [1970] 3 All E.R. 915.
26. *Cooksey and Bibby* v. *Rednall* [1949] 30 T.C. 514.
27. *Cooke* v. *Haddock* [1960] 39 T.C. 64 and *IRC* v. *Livingstone* [1926] 11 T.C. 538.
28. S.776 deals, *inter alia*, with situations arising in property development and dealing, asset stripping, the ownership of shares in property companies and inter-group transfers. Private residences are exempt (ICTA 1988 s.776 (9)).
29. These provisions have been progressively reduced over recent years.

TABLE OF CASES

TABLE OF STATUTES

APPENDICES

Tax rate tables for inheritance tax for 1989–90, 1988–89, 1987–88 and 1986–87

Transfers on death after 6 April 1989 and before 5 April 1990

Tax on transfer
Slice of cumulative

Chargeable transfer	Cumulative total	Rate
£	£	%
< 118 000	0–118 000	nil
> 118 000		40

Grossing-up of transfer (where applicable)

Net transfer	Tax payable
£	£
> 118 000	nil
> 118 000	nil + ⅔ of amount > 118 000

Transfers during lifetime after 6 April 1989 and before 5 April 1990

Tax on transfer
Slice of cumulative

Chargeable transfer	Cumulative total	Rate
£	£	%
< 118 000	0–110 000	nil
> 118 000		20

Grossing-up of transfer (where applicable)

Net transfer	Tax payable
£	£
< 118 000	nil
> 118 000	nil + ¼ of amount > 118 000

Transfers on death after 16 March 1987 and before 15 March 1988

Tax on transfer
Slice of cumulative

Chargeable transfer	Cumulative total	Rate
£	£	%
< 90 000	0– 90 000	nil
next 50 000	90 001–140 000	30
next 80 000	140 001–220 000	40
next 110 000	220 001–330 000	50
> 330 000		60

Grossing-up of transfer (where applicable)

Net transfer	Tax payable
£	£
< 90 000	nil
next 35 000	nil + $\frac{3}{7}$ of amount > 90 000
next 48 000	15 000 + $\frac{2}{3}$ of amount > 125 000
next 55 000	47 000 + 100% of amount > 173 000
> 228 000	102 000 + $\frac{3}{2}$ of amount > 228 000

Transfers during lifetime after 16 March 1987 and before
15 March 1988

Tax on transfer
Slice of cumulative

Chargeable transfer	Cumulative total	Rate
£	£	%
< 90 000	0– 90 000	nil
next 50 000	90 001–140 000	15
next 80 000	140 001–220 000	20
next 110 000	220 001–330 000	25
> 330 000		30

Grossing-up of transfer (where applicable)

Net transfer		Tax payable		
	£		£	
	< 90 000		nil	
next	42 000		nil + $^3/_{17}$	of amount > 90 000
next	64 000		7 500 + $^1/_4$	of amount > 132 000
next	82 000		25 000 + $^1/_3$	of amount > 196 500
	> 279 000		51 000 + $^3/_7$	of amount > 279 000

Transfers on death after 17 March 1986 and before 17 March 1987

Tax on transfer
Slice of cumulative

Chargeable transfer		Cumulative total	Rate
	£	£	%
	< 71 000	0– 71 000	nil
next	24 000	71 001– 95 000	30
next	34 000	95 001–129 000	35
next	35 000	129 001–164 000	40
next	42 000	164 001–206 000	45
next	51 000	206 001–257 000	50
next	60 000	257 001–317 000	55
	> 317 000		60

Grossing-up of transfer (where applicable)

Net transfer		Tax payable		
	£		£	
	< 71 000		nil	
next	16 800		nil + $^3/_7$	of amount > 71 000
next	22 100		7 200 + $^7/_{33}$	of amount > 87 800
next	21 000		19 100 + $^2/_3$	of amount > 109 900
next	23 000		33 100 + $^9/_{11}$	of amount > 130 900
next	25 500		52 000 + 100%	of amount > 154 000
next	27 000		77 500 + $1^1/_9$	of amount > 179 500
	> 206 500		110 500 + $^3/_2$	of amount > 206 500

Transfers during lifetime after
17 March 1986 and before 17 March 1987

Tax on transfer
Slice of cumulative

Chargeable transfer	Cumulative total	Rate
£	£	%
< 71 000	0– 71 000	nil
next 24 000	71 001– 95 000	15
next 34 000	95 001–129 000	17.5
next 35 000	129 001–164 000	20
next 42 000	164 001–206 000	20
next 51 000	206 001–257 000	25
next 60 000	257 001–317 000	27.5
> 317 000		

Grossing-up of transfer (where applicable)

Net transfer	Tax payable		
£	£		
< 71 000	nil		
next 20 400	nil + $3/17$	of amount >	71 000
next 28 050	3 600 + $7/33$	of amount >	91 400
next 28 000	9 550 + $1/4$	of amount >	119 450
next 32 550	16 550 + $9/31$	of amount >	147 450
next 38 250	26 000 + $1/3$	of amount >	180 000
next 43 500	38 750 + $11/29$	of amount >	218 250
> 261 750	55 250 + $3/7$	of amount >	261 750

Leases – restriction of allowable expenditure (CGTA 1979 ss.127, 129 and Sch.3)

Year	%	Year	%
50 – or more	100	25	81.100
49	99.657	24	79.622
48	99.289	23	78.055
47	98.902	22	76.399
46	98.490	21	74.635
45	98.059	20	72.770
44	97.595	19	70.791
43	97.107	18	68.697
42	96.593	17	66.470
41	96.041	16	64.116
40	95.457	15	61.617
39	94.842	14	58.971
38	94.189	13	56.167
37	93.497	12	53.191
36	92.761	11	50.038
35	91.981	10	46.695
34	91.156	9	43.154
33	90.280	8	39.999
32	89.354	7	35.414
31	88.371	6	31.195
30	87.330	5	26.722
29	86.226	4	21.983
28	85.053	3	16.959
27	83.816	2	11.629
26	82.496	1	5.983
		0	0

APPENDIX 3

Retail Price Index (FA 1982 s.3, 86–89 and Sch.13 and FA 1985 s.68 and Sch.19)

	1982	1983	1984	1985	1986	1987	1988	1989
January	78.7	82.6	86.8	91.2	96.2	100.0	103.3	111.0
February	78.8	83.0	87.2	91.9	96.6	100.4	103.7	111.8
March	79.4	83.1	87.5	92.8	96.7	100.6	104.1	112.3
April	81.0	84.3	88.6	94.8	97.7	101.8	105.8	114.3
May	81.6	84.6	89.0	95.2	97.8	101.9	106.2	115.0
June	81.9	84.8	89.2	95.2	97.8	101.9	106.6	115.4
July	81.9	85.3	89.1	95.2	97.5	101.8	106.9	115.5
August	81.9	85.7	89.9	95.5	97.8	102.1	107.9	115.8
September	81.9	86.1	90.1	95.4	98.3	102.4	108.4	116.6
October	82.3	86.4	90.7	95.6	98.5	102.9	109.5	117.5
November	82.7	86.7	91.0	95.9	99.3	103.4	110.0	118.5
December	82.5	86.9	90.9	96.0	99.6	103.3	110.3	118.8

INDEX